HOUSE OF ABUNDANCE
PUBLICATIONS

Quantum Physics Unplugged

*A Beginner's Guide to Understanding the Universe's
Greatest Mysteries - Master the Basics Through Clear
Language, Fun Examples, and Zero Complex Math*

First edition

This book was professionally typeset on Reedsy.
Find out more at reedsy.com

"Everything we call real is made of things that cannot be regarded as real."

NIELS BOHR

Contents

I   Part 1: Foundations of the Quantum Universe

1	Introduction	3
2	Simplicity: When Giants Stumble	10
3	Paradox - Light's Double Life	24
4	Ambiguity - The Probability Cloud	34

II   Mind-Bending Quantum Concepts

5	Certainty Limits - The Edge of Knowledge	43
6	Entanglement - The Invisible Thread	53
7	Transformation: Reality's Mirror	66

III   Quantum Technology and the Future

8	Innovation - Tomorrow's Technology	77
9	Medicine - Nature's Secret	86
10	Expansion - The Quantum Universe	94
11	Conclusion: Your Quantum Journey	102

Glossary: Understanding the Language of Quantum Physics   105

Recommended Bundle   113
Entangle Others in the Wonder of Quantum Physics   115

I

Part 1: Foundations of the Quantum Universe

1

Introduction

You're wielding quantum physics every time you snap a photo with your smartphone. That crisp, vibrant selfie? It exists because of quantum mechanics working inside your phone's image sensor. Yet no one ever told you that, did they? This is just one example of how quantum physics quietly runs your daily life, even though most think it's locked away in research labs or theoretical papers.

I remember the first time I tried to learn quantum physics. I was staring at a textbook filled with intimidating equations and abstract concepts, feeling my confidence drain with each incomprehensible page. Maybe you've been there, too—trying to understand concepts like "quantum superposition" or "wave-particle duality" only to feel more confused than when you started. Perhaps you've wondered why something so fundamental to our modern world seems accessible only to those with advanced degrees in physics.

Here's the thing: quantum physics isn't just hiding in your

smartphone camera. It's running your life in ways you'd never expect. When you use GPS to navigate to that new restaurant? Quantum effects must be accounted for, or you'd end up miles off course. That LED screen you may be reading this on? Quantum mechanics makes those pixels light up. Are solar panels powering more and more homes? They work because of quantum physics. Photosynthesis, the process that gives us the oxygen we breathe, relies on quantum phenomena.

But if quantum physics is fundamental to our daily lives, why does it seem impossible to understand?

I've spent years watching brilliant minds get stuck at the same roadblocks. They pick up a quantum physics book, encounter a wall of mathematical equations, and put it down, feeling defeated. They watch online lectures that zoom past crucial concepts, leaving them more confused than enlightened. They read articles that either oversimplify everything into meaningless analogies or dive so deep into technical jargon that they might as well be reading ancient Sanskrit.

Think about the last time you tried to learn about quantum physics. It could be an article about quantum computers that caught your eye or a fascinating documentary about the nature of reality. You started reading or watching with excitement, ready to unlock the mysteries of the universe. But then came the avalanche of complex terminology: wave functions, quantum entanglement, superposition, quantum tunneling. Before you knew it, that initial excitement turned to frustration, and you wondered if this knowledge was beyond your reach.

This frustration isn't just about missing out on fascinating science – it's about feeling left behind in a world racing toward a quantum revolution. Quantum computers are emerging that could revolutionize medicine, crack current encryption methods, and transform artificial intelligence. Major companies are investing billions in quantum technology. Financial institutions are preparing for quantum-safe cryptography. Healthcare companies are exploring quantum computing for drug discovery. Tech giants are developing quantum sensors that could transform everything from medical imaging to navigation systems.

The quantum future isn't coming – it's already here, transforming industries in ways that would have seemed like science fiction just a decade ago. In hospitals worldwide, medical researchers are peering into the human body with unprecedented precision, using quantum sensors to spot diseases at the molecular level before they become life-threatening. Their work promises to revolutionize early detection and treatment, potentially saving millions of lives.

Meanwhile, in the digital realm, cybersecurity experts are racing against time. They know that quantum computers will soon be able to crack today's encryption methods, so they're developing new quantum-resistant security systems to protect our digital lives. It's like rebuilding the walls of a fortress while knowing that a powerful new weapon is being developed that could breach them.

Energy companies aren't sitting idle either. They're harnessing quantum computing to solve complex optimization problems that could make our power grids more efficient and help develop the next generation of batteries. These advancements could accelerate our transition to renewable energy and help combat climate change.

In the fields and farms that feed our world, agricultural scientists are deploying quantum sensors to monitor soil conditions and crop health with extraordinary precision. These tools allow farmers to understand their land at the molecular level, optimizing water usage and reducing the need for pesticides and fertilizers.

Perhaps most surprisingly, quantum technology is transforming the world of finance. Banks and investment firms are exploring quantum algorithms that could revolutionize risk analysis and fraud detection. These systems could spot patterns and anomalies that traditional computers might miss, making our financial system more efficient and secure.

Using the revolutionary S.P.A.C.E.T.I.M.E. Method, a framework that breaks down quantum physics into nine digestible concepts: Simplicity, Paradox, Ambiguity, Certainty limits, Entanglement, Transformation, Innovation, Medicine, and Expansion, this book will transform how you understand quantum physics. No complex math. No impenetrable jargon. Just clear explanations connected to the technology you use every day.

Think of this book as your quantum physics translator. Instead of drowning you in equations, I'll show you how quantum tunneling makes your computer's transistors work. Rather than abstract theories, you'll learn how quantum entanglement powers the next generation of ultra-secure communications. Each chapter builds your understanding through real-world examples and applications you can see and touch.

The journey through this book follows the same path our minds naturally take when learning something new: we start with what we know and gradually venture into the unknown. Think of it as exploring a fascinating new city. We'll begin on familiar streets, like the technology you use daily, before wandering down increasingly exotic alleyways of quantum understanding.

You might be surprised to learn that quantum physics is already your constant companion. Every time you check your location on your phone, quantum effects make that blue dot on your map more accurate. When you step outside and feel the sun's warmth on your face, you're experiencing the result of quantum processes that began deep in the sun's core. Even the plants in your garden rely on quantum tunneling to turn sunlight into energy through photosynthesis.

As we explore deeper, you'll discover why quantum computers aren't just faster versions of regular computers but fundamentally different machines that could revolutionize everything from drug discovery to climate modeling. We'll see how quantum technology can transform medicine, reshape the financial world, and create unhackable communication networks.

By the end of our journey, you'll find yourself looking at the world differently. The smartphone in your pocket will become more than just a device – you'll understand the quantum principles that make its components work. When you hear news about quantum computing breakthroughs, you'll grasp what they mean and why they matter. You'll be able to confidently join conversations about quantum technology, separating science fact from science fiction.

Most importantly, you'll develop an intuitive feel for quantum physics without drowning in complex mathematics. You'll see how quantum mechanics shapes our present and future, from the smallest microchip to the most significant questions about the nature of reality itself. And if you choose to dive deeper into more advanced quantum concepts, you'll have a solid foundation upon which to build.

This isn't just about adding new knowledge – it's about transforming how you see the world around you. The quantum realm might seem strange initially, but by the end of this book, you'll find it as familiar as your reflection.

The best part? You don't need any prior physics knowledge. If you can use a smartphone, you can understand quantum physics. This isn't about memorizing formulas – it's about seeing the quantum world that's been around you all along. It's about understanding the principles that will shape our technological future.

This book is for the curious minds who know there's more to quantum physics than Schrödinger's famous cat. It's for professionals who want to understand how quantum computing might transform their industry. It's for parents who wish to explain how the universe works to their children. It's for anyone who's ever looked at the mysteries of the quantum world and thought, "There must be a better way to learn this."

Most importantly, this book is for you if you've ever felt that quantum physics was beyond your reach. These concepts aren't reserved for physicists and mathematicians – they belong to everyone who wants to understand the fundamental workings of our universe and the technology that will shape our future.

Welcome to quantum physics unplugged. Let's decode the universe together, one concept at a time.

2

Simplicity: When Giants Stumble

I n 1897, physicist William Thomson (Lord Kelvin) made one of science's most infamous declarations: "There is nothing new to be discovered in physics now. All that remains is more and more precise measurement."

That same year, Ernest Rutherford discovered the electron—and suddenly, the entire foundation of physics began to crack. It was as if someone had confidently declared they'd mapped every street in New York City, only to discover an entire underground network of roads that didn't follow the usual traffic rules.

The discovery of the electron wasn't just another scientific breakthrough—it was the first glimpse into a hidden world that would forever change our understanding of reality. Imagine being a master chef who's perfected every recipe in your cookbook, only to discover an entirely new type of ingredient that doesn't follow any of your cooking rules. That's precisely what happened to physicists at the turn of the 20th century. The electron, this tiny particle that helps power your smartphone and makes lightning possible, refused to behave according to the established laws of physics. It was like finding a liquid that flows upward instead of down or a ball that bounces higher each time it hits the ground.

What is Quantum Physics?

If you've ever tried learning a new language, you know that moment when you realize you can't just translate word-for-word from English. Each language has its own rules, logic, and way of expressing ideas that might not even exist in your native tongue. Quantum physics is similar—it's the language of the incredibly small and comes with seemingly bizarre grammar.

Classical physics is like watching a football game from the stands. You can see the whole field, track the ball, and predict where players will move. But quantum physics? That's like trying to watch the same game through a keyhole. Suddenly, the rules that made perfect sense from above become confusing and unpredictable. You might see a player appear in one place. Seemingly teleport to another, or somehow be in two places simultaneously.

At its core, quantum physics is the science of the tiny atoms and the even tinier particles that make them up. But calling it "the science of the very small" is like calling the Internet "a bunch of connected computers." Technically accurate, but it misses the revolutionary impact it has on our world.

The quantum world isn't just some abstract realm confined to physics laboratories—it's woven into the fabric of your daily life in ways that would seem magical if you didn't know the science behind them. Take your morning commute, for instance. You rely on quantum effects when you check your phone's GPS to avoid traffic. Without accounting for these quantum influences, your location would drift by miles, sending you to the wrong exit or down the wrong street. Those satellites orbiting high above Earth need quantum mechanics to tell you precisely where you are.

Look closer at your phone's screen. Each pixel that lights up to form text, images, and videos works because of quantum mechanics' strangest feature: quantum tunneling. Particles in your phone's display do something that should be impossible—they pass through solid barriers as if walking through walls. It's not a glitch or a malfunction; it's a fundamental part of how our universe works at its most minor scales.

The natural world performs its quantum magic every day. As you walk past a tree or tend to your garden, you witness quantum mechanics. Those green leaves aren't just pretty—they're sophisticated quantum machines. Photosynthesis, the process that keeps plants alive and provides us with oxygen to breathe, achieves its remarkable efficiency by exploiting quantum effects. It's as if nature figured out quantum mechanics billions of years before we did.

Even your own body is quantum-powered. The DNA that makes you who you are occasionally undergoes mutations—changes that drive evolution itself. These mutations often occur because of quantum tunneling, the same ghostly phenomenon that makes your phone's screen work. Above us all, the sun continues to shine because quantum tunneling allows atomic nuclei to fuse, releasing the energy that lights and warms our world. Without quantum mechanics, the sun would go dark, and life as we know it would be impossible.

Quantum physics operates at a scale so small it's hard to imagine. If you magnify a hydrogen atom until its nucleus is the size of an orange, its electron would be a speck about 2.5 miles away. Yet, understanding this realm has led to technologies that define our modern world.

When Newton's Laws Weren't Enough

We all intuitively know Newton's laws. Throw a ball up; it comes down. Push a shopping cart; it moves forward. These laws work brilliantly for the world we can see. They're why engineers can build bridges that don't collapse and why NASA can land rovers on Mars.

Newton's laws elegantly explain the world around us. Watch a car roll to a stop at a red light, and you'll see his First Law of Motion in action—the car wants to keep moving forward until the brakes overcome its inertia. Every morning commute is a testament to Newton's genius.

When you watch a rocket launch on TV, you see Newton's Third Law performing its cosmic ballet. The massive thrust of flame shooting downward isn't just for show—it's the other half of the action-reaction pair that hurls spacecraft into the heavens. Each pound of thrust pushing down creates an equal force pushing the rocket up, carrying human dreams into space.

Step into a pool hall, and you'll find Newton's laws governing every collision and ricochet. The satisfying crack of a perfect break shot demonstrates the conservation of momentum as energy flows from the cue ball through the neat triangle of balls, sending them spinning across the green felt in predictable patterns.

Even something as thrilling as skydiving follows Newton's orderly rules. As a diver plummets through the air, gravity pulls it downward until air resistance builds to match it perfectly. At this equilibrium point, they reach terminal velocity—a perfect balance of forces Newton would have understood completely.

And far above us, planets dance around the sun in endless orbital loops, guided by the same law of universal gravitation that brings a tossed ball back to Earth. From the most miniature game of catch to the grandest cosmic ballet, Newton's laws seemed to explain everything—until we looked much closer.

But something strange happened when scientists tried to apply these same laws to the atomic world. Imagine you're playing pool. Newton's laws tell you exactly how the balls will move when collating. Now, shrink those pool balls to the size of atoms, and suddenly, nothing makes sense anymore. The balls might tunnel through each other, appear in two places at once, or somehow affect each other from across the table without touching.

Here are three dramatic examples where classical physics completely broke down:

1. The Case of the Impossible Atom

According to Newton's laws, electrons orbiting an atom's nucleus should spiral inward and crash, like a satellite eventually falling to Earth. The math was straightforward: atoms shouldn't exist for more than a fraction of a second. Yet here you are, reading this book, made of atoms that have been stable for billions of years. Classical physics couldn't explain why you weren't imploding this very moment.

2. The Light That Shouldn't Work

Light hitting certain metals can knock electrons loose—this is called the photoelectric effect. Classical physics predicted that brighter light should give electrons more energy. Instead, the power depends on the color (frequency) of the light, not its brightness. It's like saying the temperature of your shower depends on the color of your shower head, not how far you turn the handle. This made no sense under classical physics.

3. The Hot Object Paradox

Classical physics insisted that hot objects emit radiation at all frequencies equally, releasing an infinite amount of energy. A simple hot cup of coffee should release enough power to destroy the universe. Spoiler alert: your morning coffee has yet to cause the apocalypse.

The Ultraviolet Catastrophe: When Physics Predicted the Impossible

The "hot object paradox" became known as the ultraviolet catastrophe, and it was more than just an intriguing puzzle—it was a crisis that threatened to undermine all of physics. Imagine you're using a map app that's always been reliable. Still, one day, it tells you to drive your car off a cliff because that's technically the shortest route. Would you trust that app again? Would you question everything it had told you before?

This was the situation physicists faced. Their trusted theories, which worked perfectly for centuries, suddenly made impossible predictions. It was like having a calculator that correctly solved every math problem except those involving the number 7, where it would suddenly claim that $7 + 7 =$ infinity.

This was so important because everything in the universe emits radiation based on its temperature. The sun, your body, that cup of coffee on your desk—they're all constantly emitting energy. Classical physics had a formula for this and worked well for low frequencies (like radio waves). But when applied to higher frequencies (like ultraviolet light), the formula predicted that any warm object should emit infinite energy.

Please take a moment to consider the absurdity of what classical physics was telling us. That simple cup of coffee sitting on your desk—the one you're sipping while reading this book—should, according to the old laws of physics, be releasing enough energy to vaporize our entire planet. You're having breakfast next to a doomsday device every morning, yet here you are, peacefully stirring your coffee.

Flip on a light switch, and classical physics insists you've just triggered an apocalyptic event. That humble bulb in your lamp should flood the universe with infinite energy the instant it illuminates. Yet, your electricity bill remains finite, and your living room hasn't become a cosmic catastrophe.

The sun, that familiar yellow disk in the sky that's faithfully warmed our planet for billions of years, should have destroyed our entire solar system long before the first dinosaurs walked the Earth. According to classical physics, it should have unleashed an infinite inferno that would have made a supernova look like a sparkler.

Even more bizarre, the warmth of your body—that modest 98.6 degrees Fahrenheit—should be enough to trigger a catastrophe of cosmic proportions. Every living thing, from bacteria to blue whales, should be a walking astronomical disaster. Yet here we all are, our body heat doing nothing more dramatic than fogging windows on cold mornings.

Yet none of this happens. Your coffee typically gets cold, light bulbs work fine, and we're all still here. Something had to be fundamentally wrong with the classical understanding of physics.

Why Tiny Things Behave Differently

Think about throwing a tennis ball. Now imagine throwing a soap bubble. The tennis ball follows a nice, predictable path that Newton's laws describe perfectly. The soap bubble? It weaves and bobs, affected by the slightest air current. The smaller something is, the more different forces come into play.

Now, shrink to the size of an electron—about 1/100,000th the size of an atom, which is already incredibly tiny. At this scale, things get weird. Weird. An electron isn't like a small tennis ball or a soap bubble. It's more like nothing we experience in our everyday world.

Let me give you some mind-bending examples of how the quantum world differs from our everyday experience:

1. Location Uncertainty

In our world, your car is either in the garage or not. But an electron can be in multiple places simultaneously—imagine your car being in your garage and your neighbor's garage simultaneously until someone looks to check where it is.

2. Quantum Tunneling

It's as if you threw a ball at a brick wall, and sometimes the ball would pass right through to the other side, completely intact. This isn't science fiction—it's how solar fusion works and how some types of computer memory store information.

3. Action at a Distance

Imagine having two coins that always land on opposite sides, even on different continents. When you flip one, the other instantly responds. Einstein called this "spooky action at a distance" a natural phenomenon called quantum entanglement.

4. The Observer Effect

In our world, looking at something doesn't change it. But in the quantum world, the mere act of observation can alter the outcome of an experiment. It's like having a friend who always acts differently when they know you're watching.

The Birth of Quantum Theory: A Detective Story

The story of quantum physics reads like a scientific thriller, complete with unlikely heroes, dramatic revelations, and philosophical debates that continue to this day. Let's meet our cast of characters and see how they unveiled the quantum world.

The First Mystery: Black Body Radiation (1859-1900)

Gustav Kirchhoff started our story by studying how objects emit radiation when heated. It seemed simple enough—heat something, and it glows. But the mathematics produced nonsensical results, predicting infinite energy from finite objects.

Enter Max Planck, a conservative physicist who made a radical proposal out of desperation. Imagine if you could only withdraw money from your bank account in $10 increments—no amounts in between. Planck suggested energy worked the same way, coming only in discrete packets he called "quanta." He didn't believe this was true; he thought it was just a mathematical trick to make the equations work.

In 1905, a young patent clerk named Albert Einstein did something audacious with Planck's revolutionary idea. Where others saw a mathematical trick, Einstein saw a fundamental truth about light. He proposed that light wasn't just a wave flowing through space like ripples on a pond—it was also a stream of tiny particles, later called photons. This wasn't just splitting hairs; it solved mysteries that had puzzled scientists for years.

Scientists had long been baffled by how metals behaved when exposed to light. Shine a bright red light on certain metals, and nothing happens. But shine even a dim violet light, and electrons suddenly burst forth from the metal's surface. It made no sense—until Einstein explained that each photon carried a specific amount of energy determined by its color. Violet photons packed enough punch to knock electrons loose; red ones didn't, no matter how many you used. It was like discovering that throwing a single golf ball could break a window while throwing a thousand tennis balls wouldn't leave a scratch.

Eight years later, Niels Bohr took quantum ideas in an even more radical direction. He turned his attention to the atom—specifically, the puzzle of why atoms were stable. According to classical physics, electrons should quickly spiral into the nucleus, causing all matter to collapse. Yet here we are, solid as ever.

Bohr's insight was revolutionary: electrons couldn't orbit just anywhere. They were confined to specific energy levels, like athletes assigned to particular seats in a stadium, with no standing room allowed. When electrons jumped between these levels, they absorbed or emitted exact amounts of energy—explaining why heated elements always produced specific light colors. It was nature's light show, precisely choreographed by quantum rules.

Then came Werner Heisenberg in 1927 with perhaps the most profound insight. He discovered that the quantum world enforces a cosmic speed limit on knowledge. Want to know exactly where an electron is? OK, but you'll lose track of its speed. Want to measure its speed precisely? You'll have to give up knowing its exact location. This wasn't about having better tools or being more careful—it was a fundamental limit written into the fabric of reality.

This strange new quantum world needed interpretation, and in 1927, the greatest minds in physics gathered at the Solvay Conference to make sense of it all. What emerged was the Copenhagen Interpretation, a radical new way of thinking about reality. It suggested that quantum systems exist in multiple states simultaneously, like a cosmic game of possibility, until we measure them. The act of measurement forces reality to "choose" one definite outcome from all these possibilities.

Einstein, who had started this quantum revolution, couldn't accept its strange implications. "God does not play dice with the universe," he famously objected. With characteristic dry wit, Bohr reportedly replied, "Einstein, stop telling God what to do." Their debate would continue for decades, but the quantum die had been cast. Physics would never be the same.

All this talk of quantum mechanics might seem abstract—like something trapped in the realm of theoretical physics—but look no further than your pocket to find it at work. That smartphone you carry isn't just a triumph of engineering; it's a quantum mechanical marvel. Every time you tap its screen, billions of transistors exploit quantum tunneling to process your commands. The bright, crisp display that lights up to show your messages? It works because we understand how electrons jump between precise energy levels in quantum systems. Even the solar panels charging your wireless earbuds harness Einstein's explanation of the photoelectric effect.

Step into any modern hospital, and you'll find quantum mechanics saving lives. MRI machines peer inside our bodies by choreographing the quantum dance of atomic nuclei. PET scanners use antimatter—positrons, the electron's mysterious twin—to spot diseases before they become visible through other means. Quantum dots light up cancer cells that would otherwise hide from view while precisely controlled laser beams, guided by quantum principles, perform surgeries with incredible precision.

Our connected world itself rests on quantum foundations. Every time you stream a movie or send a text, that information races through fiber optic cables as quantum particles of light. The GPS that guides you through city streets only works because it accounts for quantum effects on satellite clocks. These corrections are necessary for you to be allowed to your intended destination. And now, quantum cryptography promises to create truly unbreakable codes protected by the fundamental laws of physics.

But what we see today is just the beginning. We're standing on the threshold of a second quantum revolution. Quantum computers are beginning to solve problems that would take classical computers longer than the universe's age to crack. Quantum sensors promise to detect diseases at the molecular level, potentially spotting health issues years before symptoms appear. The quantum internet could make today's security concerns obsolete, creating communication networks protected by the strange rules of quantum entanglement.

In research labs worldwide, scientists are pushing the boundaries of what's possible. They're creating new quantum materials with properties that defy classical physics. They're developing sensitive quantum sensors that can detect a single atom out of place. They're building quantum computers that don't just process information differently—they harness the weirdness of the quantum world to perform computations that classical computers can't even approach.

And this is just the beginning. The quantum technologies we're developing today could transform every field, from medicine to climate science, financial modeling to artificial intelligence. We're learning not just to understand the quantum world but to shape it to our needs.

As we move forward, prepare yourself for something even more mind-bending. In the next chapter, we'll explore one of the most famous quantum experiments ever performed: the double-slit experiment. You'll see how single particles can exist in multiple places at once and how looking at them fundamentally changes their behavior. If you think what we've covered so far is strange, as Niels Bohr famously said, "If quantum mechanics hasn't profoundly shocked you, you haven't understood it yet." But don't worry—understanding why it's shocking is the first step to understanding how it works.

The quantum future isn't coming—it's already here, stranger and more wonderful than we could have imagined.

3

Paradox - Light's Double Life

W hat if I told you that light can simultaneously behave like a gentle ripple in a pond and a speeding bullet? As physicist Niels Bohr once said, "Anyone who is not shocked by quantum theory has not understood it." The story of light's double life is one of the strangest—and most important—discoveries in science, and it's about to blow your mind.

Imagine watching a superhero movie in which the main character can be as solid as steel and pass through walls like a ghost. Sounds impossible, right? Yet, this is exactly what light does in the real world. It's both a wave, spreading out like ripples in a pond, and a particle, behaving like tiny bullets. This isn't science fiction—it's the reality that powers everything from your smartphone's camera to the internet itself.

The Double-Slit Experiment: The Most Beautiful Experiment in Physics

Imagine being in London in 1801, watching a young scientist named Thomas Young set up what would become known as "the most beautiful experiment in physics." The setup was deceptively simple: a beam of light, two narrow slits cut into a card, and a screen to catch the light. But what this humble arrangement revealed would shake the foundations of physics and unveil one of nature's most profound mysteries.

To understand why this experiment was so revolutionary, picture throwing tennis balls at a wall with two narrow openings. Common sense tells us you'd find two neat piles of balls on the other side, each pile aligned with its respective slit. That's how particles should behave—they pick and stick to a path.

Now imagine dropping two pebbles into a calm pond. The ripples spread out, and something magical happens when they encounter a barrier with two gaps. The waves passing through these gaps spread out again, meeting and interacting. Where wave peaks meet, they create more significant peaks; where a peak meets a trough, they cancel each other out. This creates a distinctive pattern of alternating strong and weak ripples.

When Young shone his light through the slits, he saw something remarkable—bands of light and darkness, just like the interference pattern of water waves. It seemed to settle a centuries-old debate: light must be a wave, flowing through space like ripples on a cosmic pond.

But the story doesn't end there. Fast forward to the modern era, when scientists developed equipment sensitive enough to detect individual particles of light—photons. They decided to repeat Young's experiment, but with a twist: they would fire these photons one at a time, like shooting individual bullets through the slits. Indeed, they thought, each photon would pick one slit or the other, eventually building up two simple stripes of light.

What they saw instead defied all logic. Even firing single photons, one at a time, slowly built up the same interference pattern that Young had seen. It was as if each particle of light was somehow going through both slits simultaneously, interfering with itself. Imagine throwing a single tennis ball and somehow having it go through both holes at once—impossible in our everyday world but routine in the quantum realm.

The mystery deepened when scientists tried to catch these quantum acrobats in the act. They set up detectors to watch which slit each photon passed through. But the moment they did, the interference pattern vanished, replaced by the two simple stripes they had initially expected. It was like discovering you have a friend who performs impressive dance moves alone but instantly freezes into ordinary walking the moment you try to film them.

This isn't just some curious laboratory phenomenon. This same quantum wizardry—particles being waves, existing in multiple places at once, and changing behavior when observed—makes your digital camera work. It's what enables fiber optic cables to carry vast data globally. The "impossible" has become not just possible but practical.

Even Einstein, who had revolutionized physics with his theories of relativity, struggled to accept these implications. But nature has little regard for what we find reasonable. In the quantum world, the impossible isn't just possible—it's mandatory.

Einstein and the Photoelectric Effect: Light Gets Particle-ar

While Young's dancing light waves captivated the scientific world, a different experiment quietly challenged everything physicists thought they knew about light. It started with a simple observation: when light strikes metal, it can knock electrons loose from their atomic bonds. Scientists called this the photoelectric effect, though it might as well have been called "light's particle punch."

According to the physics of the time, this effect should have been straightforward. Shine any light color on metal long enough; eventually, you'd provide enough energy to free some electrons. It was like pushing a child on a swing—even gentle pushes, given enough time, should eventually get the swing moving.

But nature had other ideas. When scientists experimented, they found something bizarre. A dim ultraviolet light would instantly send electrons flying from the metal surface. In contrast, even the brightest red light wouldn't budge a single electron, no matter how long it shone. Making the light brighter only produced more freed electrons, not faster-moving ones. It was as if the universe was playing by rules no one understood.

Enter Albert Einstein, who proposed something revolutionary: What if light wasn't just a wave but also came in tiny packets of energy, like miniature bullets? These packets—which we now call photons—would carry different amounts of energy depending on their color. Ultraviolet photons packed a powerful punch, while red photons were comparative weaklings.

Suddenly, the photoelectric effect made perfect sense. Each ultraviolet photon carried enough energy to knock an electron free immediately like a well-aimed golf ball breaking a piggy bank. Red photons, no matter how many you used, were like throwing ping pong balls at the bank—they didn't have enough individual energy to break anything loose. Brighter light meant more photons, like throwing more golf balls, but each electron was still only hit by one photon at a time.

This discovery, which would earn Einstein the Nobel Prize, revealed light as nature's ultimate shape-shifter. Like Mystique from X-Men morphing between different forms, light could act as either a wave or particle, depending on the situation. Light spreads out like a wave when traveling through space, exploring all possible paths. But when it interacts with matter—striking a metal surface or hitting your digital camera's sensor—it behaves like a particle, delivering its energy in discrete chunks.

This dual nature isn't a flaw or a contradiction—it's fundamental to how our universe works at its deepest level. We see it in action every time we use the internet. In fiber optic cables, light travels as waves, spreading through glass fibers with minimal loss. But when that light reaches its destination, photon detectors catch individual particles of light, converting them into the electrical signals that become your streaming videos, social media posts, and email.

Light, it turns out, isn't confused at all—we were the confused ones, trying to force it into our limited categories of "wave" or "particle." Nature refused to be so easily labeled, teaching us that reality is far richer and stranger than we had imagined.

Tony Stark Would Be Impressed: Modern Technologies That Use Light's Dual Nature

Remember the Arc Reactor from Iron Man? While Tony Stark's power source remains in science fiction, the fundamental technologies that harness light's dual nature are equally impressive. They're transforming our world in ways that make even Iron Man envious.

Take solar panels, for instance. These sleek energy collectors work because we understand both faces of light. Photons that strike the panel's surface behave like particles, knocking electrons loose through the photoelectric effect. Yet engineers design these panels by treating sunlight as a wave, carefully tuning them to capture specific wavelengths for maximum efficiency. This marriage of light's dual nature has made solar power a clean energy source increasingly viable.

Your digital camera tells a similar story. When you take a photo, light waves focus through your camera's lens system, bending and converging like ripples in a pond. But when this light hits the sensor, individual photons strike like tiny bullets, triggering precise electrical signals in each pixel. Your selfies exist because we've learned to manipulate both aspects of light's personality.

Lasers represent the most dramatic mastery of light's dual

nature. By synchronizing light waves and controlling the flow of photons, we create intense beams that can cut through steel, perform delicate eye surgery, or read the microscopic pits in a Blu-ray disc. The same technology guides precision manufacturing, measures distances in construction, and enables fiber optic networks that form the backbone of the internet.

Speaking of displays, your phone's screen is a quantum light show. When electrons in the LED display jump between different energy states, they release precisely controlled photons of specific colors. These quantum leaps create every image you see, from your favorite social media feed to streaming videos.

The Quantum Revolution in Your Pocket

Your smartphone isn't just a communications device—it's a masterpiece of quantum engineering. The camera system alone brilliantly demonstrates this: light waves focus through precisely engineered lenses before individual photons trigger quantum effects in the sensor. The image processing software even accounts for quantum noise to produce clearer pictures.

The screen technology is equally sophisticated. Quantum dots—tiny semiconductor particles—create vivid colors by controlling how electrons jump between energy levels. LED backlights use similar quantum transitions to illuminate the display, while sophisticated light guides ensure even brightness across the screen.

In newer phones, even facial recognition relies on quantum effects. Infrared photons map your face's contours, quantum sensors detect the reflected light, and neural networks process this quantum information to verify your identity. Even your GPS location depends on quantum precision—satellite atomic clocks must account for both quantum and relativistic effects to give you accurate directions to the nearest coffee shop.

Beyond Consumer Tech: Medical and Scientific Applications

Light's quantum nature is improving our gadgets and revolutionizing medicine and scientific research. In hospitals, MRI machines peer inside our bodies by manipulating the quantum properties of atoms. PET scanners track quantum interactions to reveal metabolic processes. At the same time, X-ray machines precisely control photon energy to create detailed images while minimizing radiation exposure.

In research laboratories, quantum microscopes achieve the seemingly impossible, revealing individual atoms. Laser spectroscopy analyzes materials by examining their interaction with light at the quantum level. Perhaps most impressively, quantum sensors can now detect gravitational waves—ripples in spacetime—by measuring quantum effects in laser beams.

The Sun's Quantum Dance

Even our sun, that ancient nuclear furnace in the sky, depends on quantum effects for its existence. Nuclear fusion in the sun's core is only possible because of quantum tunneling—particles doing something classical physics says is impossible. Once created, photons begin an epic journey, taking thousands of years to reach the surface as they bounce through the sun's interior. When they finally emerge, light's wave-particle nature helps create everything from the sun's visible surface to our blue sky.

The future looks even brighter as we develop new quantum technologies. Quantum computers promise to solve problems beyond classical computers' reach. Quantum encryption could secure communications. Ultra-precise quantum measurements might reveal new phenomena we haven't even imagined yet.

Deeper Into the Quantum Realm

If light's ability to be both a wave and a particle seems strange, it's only the beginning. In the next chapter, we'll dive into one of quantum physics' most profound principles: uncertainty. You'll discover why particles don't have defined locations until we measure them and how this fundamental fuzziness of reality leads to remarkable technologies.

Remember: Your phone works because of quantum physics. So do your TV, LED lights, and car's GPS. The sun shines because of quantum tunneling. MRI machines see inside you using quantum effects. Solar panels make electricity through the photoelectric effect. Computer chips process this text using quantum mechanics.

That's fantastic about quantum physics—it's not just equations on a chalkboard. Science makes our modern world possible, and it's working its quantum magic in your pocket right now.

As we venture deeper into quantum physics, keep in mind that these "impossible" behaviors of light were just the beginning. Nature has many more surprises in store, and understanding them has given us technological superpowers that would make even Tony Stark jealous.

4

Ambiguity - The Probability Cloud

"Everything we call real is made of things that cannot be regarded as real." - Niels Bohr.

Let that sink in for a moment. One of the greatest physicists ever told us that reality itself isn't exactly... real. At least, not in the way we usually think about it. If that sounds mind-bending, you're in good company. Even Einstein struggled with this idea.

Imagine you're watching a coin spinning on a table. Before it lands, is it heads or tails? While it's spinning, it's somehow both and neither simultaneously. Now, shrink that coin down to the size of an atom, and you're beginning to grasp one of quantum physics' most profound ideas: at its most miniature scale, nature deals in possibilities rather than certainties.

Wave Functions: The Crystal Ball of Quantum Physics

Let's start with something simple: Where is an electron in an atom? You might remember pictures of electrons orbiting the nucleus like tiny planets around a sun in your school days. It's a nice image, but it needs to be corrected.

In quantum physics, we can't say exactly where an electron is. Instead, we use a wave function—a mathematical description of where it might be. Think of it like your weather app predicting tomorrow's forecast. When it shows a 70% chance of rain, it's not saying it will rain; it's just that it's likely. A wave function works similarly, telling us the probability of finding an electron in different locations.

Consider your local mall on a Saturday afternoon. Suppose someone asks where your teenage child is. In that case, there's a high probability they're at the food court, a decent chance they're at the gaming store, and a slim possibility they're doing homework in the library. A wave function describes an electron's location similarly—but with one crucial difference: until you look (measure its position), the electron is somehow in all those places simultaneously.

Welcome to the Probability Cloud

Forget everything you learned about electrons orbiting atoms like planets around the sun. The reality is far more fascinating. Instead of following fixed paths, electrons exist in "probability clouds" or orbitals. These aren't actual clouds you can see but mathematical descriptions of where an electron will likely be found at any given moment.

Think of it like predicting where a bee might be in a garden. You can't know its exact location, but it will likely be near flowers—the more attractive the flower, the higher the probability of finding the bee there. Until you spot the bee, it's as if it's visiting all the flowers at once—at least in the quantum world.

This isn't just theoretical musing—it's how atoms work. The shapes of these probability clouds determine how atoms bond, which determines the properties of every material around us. This understanding has revolutionized technology, leading to the development of semiconductors that power your phone, LED lights that illuminate your home, and solar panels that convert sunlight into electricity.

Schrödinger's Cat: The Most Famous Feline in Physics

Now we come to physics' most famous thought experiment: Schrödinger's Cat. Erwin Schrödinger devised this scenario to show how weird quantum mechanics becomes when applied to everyday objects.

Picture this: a cat is placed in a sealed box with a clever but diabolical device. Inside this box is a vial of poison, a radioactive atom, and a mechanism that will break the vial if the atom decays. The radioactive atom has a 50% chance of decaying within an hour. If it decays, the poison is released, and the cat dies. If it doesn't decay, the cat lives.

According to quantum mechanics, something extraordinary happens before we open the box. The radioactive atom exists in a superposition of both decayed and not decayed states. Consequently, the poison is both released and not released, and the cat is—impossibly—both alive and dead at the same time. Only when we open the box and observe the system does this superposition "collapse" into one definite outcome.

Of course, Schrödinger wasn't suggesting we put cats in boxes. He created this thought experiment to highlight the absurdity of applying quantum mechanical principles to everyday objects. The cat paradox illustrates a fundamental question in quantum physics: Where exactly does the quantum world end and our classical world begin? Why don't we see larger objects like cats in multiple states simultaneously?

Why Quantum Strangeness Makes Perfect Sense (Sort Of)

Despite its apparent strangeness, quantum mechanics is the most successful scientific theory ever developed. Every prediction it's made has been confirmed by experiments, and every technology based on its principles works exactly as expected. It's like having a foreign phrasebook that sounds nonsensical when read aloud but somehow gives you perfect directions every time you use it.

Take your smartphone's processor, for instance. Its billions of transistors work because we understand how electrons move through probability clouds. MRI machines create detailed images of your body by manipulating the quantum properties of atoms. The GPS in your car knows your location precisely because it accounts for quantum effects in satellite signals.

The fact that quantum mechanics seems strange to us isn't a flaw in the theory—it's a limitation of our human experience. We evolved to survive in a world of large objects where quantum effects average out. Expecting quantum physics to match our everyday intuition is like expecting a fish to understand the concept of drought.

Quantum Superposition: More Than Just a Parlor Trick

The idea that particles can exist in multiple states simultaneously might seem like a curious oddity. Still, it's becoming the foundation of a technological revolution. Quantum computers, for instance, harness this very principle to perform calculations that would be impossible for traditional computers.

While a classical computer bit must be either 0 or 1, a quantum bit (qubit) can simultaneously be in a superposition of both states. You could complete a multiple-choice test by choosing all answers simultaneously, then get the right one when it's time to turn in your paper. This isn't science fiction—companies like IBM, Google, and Microsoft are already building quantum computers based on these principles.

The Cloud Computing You Didn't Know About

The electron probability clouds we discussed earlier don't just help us understand atoms—they're essential to modern chemistry and materials science. When scientists develop new materials, they use quantum mechanics to predict how electron clouds will interact.

This has led to breakthroughs in:

- Solar cell efficiency
- Battery technology
- Drug development
- Clean energy catalysts
- Advanced electronics

Every time you use a touchscreen, you interact with materials designed using our understanding of electron probability clouds. This same quantum foundation will emerge from the next generation of flexible electronics, quantum sensors, and clean energy technologies.

Living in a Quantum World

Despite its microscopic scale, quantum probability affects our daily lives in countless ways. The sun only shines because quantum tunneling allows nuclear fusion to occur. Plant photosynthesis uses quantum coherence to convert sunlight into energy more efficiently than any human-made solar cell. Even our sense of smell might rely on quantum effects.

Understanding quantum probability has deepened our knowledge of nature and given us technological superpowers our grandparents couldn't have imagined. Every digital photo you take, every LED light you switch on, and every semiconductor device you use works because we've learned to harness the quantum realm.

The Ultimate Speed Limit

We've seen how particles exist in multiple states until observed, how electrons live in probability clouds rather than fixed orbits, and how these strange properties enable modern technology. However, we must discuss another fundamental limit in quantum physics that Einstein himself struggled to accept.

In the next chapter, we'll explore Heisenberg's Uncertainty Principle, which limits what we know about quantum systems. We'll discover why we can never simultaneously know where a particle is and where it's going with perfect precision. This isn't a limitation of our measuring devices—it's a fundamental reality feature.

Remember, the quantum world is still being determined because we lack precise measurements. It's uncertain because uncertainty is woven into the fabric of reality. As we'll see, this inherent fuzziness of nature isn't a bug—it's a feature that makes our universe possible.

The journey into quantum ambiguity might seem strange, but it's leading us to a more profound truth about nature. As Niels Bohr said, "If quantum mechanics hasn't profoundly shocked you, you haven't understood it yet." But understanding this shock is the first step toward appreciating our remarkable world—a world where probability clouds dance in every atom of your body, and uncertainty becomes a certainty.

II

Mind-Bending Quantum Concepts

5

Certainty Limits - The Edge of Knowledge

"Not only is the Universe stranger than we think, it is stranger than we can think." - Werner Heisenberg.

These words weren't just philosophical musing—they came from the physicist who discovered one of nature's most profound secrets: there are fundamental limits to what we can know about the universe. Not because our instruments aren't precise enough or because we're not smart enough, but because uncertainty is woven into the very fabric of reality.

When Certainty Crumbled

In 1925, a young Werner Heisenberg retreated to a windswept island in the North Sea, seeking relief from his hay fever. What he discovered there would shake the foundations of physics more profoundly than anything since Newton. Working late into the night, fueled by coffee and confusion, Heisenberg realized something revolutionary: the more precisely you measure specific properties of a particle, the less precisely you can know others.

This wasn't just another scientific discovery—it was a fundamental limit on knowledge itself. Einstein, who had revolutionized physics with relativity, found this idea deeply troubling. "God does not play dice," he famously objected. But Heisenberg's uncertainty principle wasn't about God or dice—it was about the nature of reality itself.

The Pool Hall That Changed Physics

Imagine you're playing pool. The green felt stretches before you, dotted with perfectly spherical balls. In this familiar world, everything follows common sense rules. You can know exactly where each ball is and how fast it moves. You can calculate perfect shots because classical physics gives you all the necessary information.

Now imagine shrinking to the size of an atom, cue stick in hand, trying to play pool with electrons instead of balls. Everything you know about pools breaks down. Try to measure where your electron "ball" is, and you lose track of its speed. Focus on its speed, and its location becomes a blur. This isn't because the electron pool is tough—it's because the universe prevents you from knowing both properties simultaneously.

Think about a game of pool for a moment. It's a perfect demonstration of classical physics in action. You can watch the white cue ball roll across the green felt, tracking its position at every instant. You can measure how fast it's moving, calculate the perfect angle for your shot, and predict where each ball will end up after the collision. Every aspect of the game follows neat, predictable rules. Advanced players can even plan multiple shots, knowing that the physics of billiards will faithfully execute their strategic vision.

This predictability is what makes Pool such a satisfying game. Whether you're calling your shot on the eight ball or setting up a complex bank shot, you're relying on the clockwork nature of classical physics. Each click of the ball against the ball, each rolling trajectory, and each gentle spin follows laws so reliable you could set your watch by them. It's a world where uncertainty has no place, where cause and effect play out like a well-rehearsed dance.

But in the quantum world, we can never have complete information. The more precisely we measure position, the more uncertain momentum becomes, and vice versa. It's not a matter of building better-measuring tools or being more careful—it's a fundamental limit written into the laws of physics.

The Camera That Couldn't

To understand why this happens, look at something familiar: photographing a moving object. When you try to capture a runner sprinting past, you face a choice:

You can use a fast shutter speed to freeze the motion, getting a crystal-clear image of where the runner is. But that sharp image tells you nothing about how fast they move—they could be running or standing still. The position is specific, but the velocity is lost.

Or you can use a slow shutter speed, creating that artistic blur that shows motion. Now, you can tell precisely how fast they're moving by measuring the length of the blur. Still, you need precise information about their location at any specific moment. The velocity is captured, but the position becomes uncertain.

This photography analogy isn't perfect, but it hints at something profound about the quantum world. In quantum mechanics, this trade-off isn't just a limitation of our cameras—it's a fundamental property of reality. The universe works like a cosmic camera that can never get a perfect position shot and a perfect motion shot of its minor components.

Why We Can't Know Everything

The reason for this uncertainty goes deeper than just measurement problems. When scientists try to measure an electron's position, they must bounce light off it. But light isn't just illumination—it carries energy and momentum. The very act of measurement changes what's being measured.

Imagine figuring out where a soap bubble is by throwing ping-pong balls at it. Sure, when a ping-pong ball hits the bubble, you know where the bubble was—but you've also popped it! The quantum world is similarly delicate. Every measurement interaction changes the system being measured.

This is more than a technical problem we might solve with better instruments someday. Heisenberg proved that this uncertainty is built into the mathematical structure of quantum mechanics. It's not that we're not clever enough to measure both properties—it's that these properties cannot simultaneously have definite values.

When Einstein Was Wrong

This revelation sparked one of the most famous scientific debates in history. Einstein, troubled by this inherent uncertainty, spent years trying to prove Heisenberg wrong. He devised increasingly clever thought experiments, attempting to show ways to measure both position and momentum precisely.

However, Heisenberg and his colleagues found flaws in Einstein's reasoning each time. The uncertainty principle held firm. Eventually, Einstein's attempts to disprove quantum uncertainty led to even deeper insights into the quantum nature of reality.

The Universe's Fuzzy Edge

This principle is revolutionary because it reveals a fundamental fuzziness in nature. Before quantum mechanics, scientists believed particles were like tiny billiard balls—small but with definite positions and velocities at all times. The uncertainty principle showed this view was wrong.

Instead of definite properties, particles exist in states of possibility until measured. It's not that the electron is in a specific place. Still, we don't know where—it doesn't have a definite position until we measure it. The same goes for its momentum.

Nature's Impossible Trick: Quantum Tunneling

When you think quantum mechanics couldn't get any stranger, consider this: particles regularly do something that should be impossible—they pass through solid barriers. Physicists call this quantum tunneling, and it's not just a curious oddity. Without it, the sun wouldn't shine, and life wouldn't exist.

Imagine rolling a ball up a hill. If it doesn't have enough energy to reach the top, it rolls back down—every time. That's classical physics. But in the quantum world, something extraordinary happens. Sometimes, particles reach the other side of the "hill" even when they don't have enough energy to climb it. It's as if you threw a tennis ball at a brick wall, and occasionally, it would pass right through, emerging unchanged on the other side.

This isn't magic—it's pure quantum mechanics. Remember the uncertainty principle we just discussed? Because a particle's position is uncertain, there's always a tiny chance it will appear on the other side of a barrier. The particle didn't break through or climb over—it "tunneled" through a region it couldn't access.

The Sun's Quantum Secret

Here's something remarkable: every sunbeam you've ever felt exists because of quantum tunneling. At the sun's core, hydrogen nuclei must get extremely close to fuse into helium. The problem is that these nuclei are positively charged, repelling each other violently. Classical physics says they could never get close enough to fuse—the energy barrier is too high.

But quantum tunneling saves the day. Because of uncertainty and tunneling, there's a tiny chance for these nuclei to appear close enough to fuse. This chance is so tiny that any particular pair of nuclei might wait billions of years to fuse. But with the astronomical number of nuclei in the sun, enough tunneling events happen every second to power all life on Earth.

Think about that moment: the light warming your skin now exists because particles did something that classical physics says is impossible.

Tunneling in Your Pocket

You don't need to look to the stars to find quantum tunneling—it's happening right now in your pocket. Every smartphone contains billions of transistors, many of which rely on quantum tunneling to function. Flash memory, which stores your photos and messages, works because electrons tunnel through fragile insulating barriers.

Modern electronics increasingly depend on this quantum "impossibility":

Scanning Tunneling Microscopes (STM) use quantum tunneling to create images of individual atoms. When the microscope's tip gets close enough to a surface, electrons tunnel between them; scientists can map surfaces with atomic precision by measuring these tunneling currents.

Your phone's processor contains tiny transistors, so quantum tunneling becomes both a challenge and a tool. Engineers must carefully control where tunneling happens and where it doesn't, turning quantum weirdness into practical computing power.

When Barriers Aren't Barriers

The implications of quantum tunneling go far beyond technology. It forces us to reconsider what we mean by a "barrier." In the classical world, barriers are absolute—either you have enough energy to overcome them, or you don't. But quantum mechanics reveals that barriers are more like probability filters. Usually, they keep things out, but there's always a chance of tunneling through.

This probabilistic nature of barriers appears throughout nature:

In biology, enzymes use quantum tunneling to speed up chemical reactions that would be impossibly slow otherwise. When hydrogen atoms need to transfer between molecules, they often tunnel through energy barriers rather than climb over them.

Some types of radioactive decay only happen because of quantum tunneling. The particles inside unstable nuclei occasionally tunnel out, leading to the steady tick of radioactive decay that helps us date ancient artifacts and power nuclear medicine.

Even the simple act of water molecules forming and breaking bonds involves quantum tunneling of hydrogen atoms. The quantum world is constantly at work in the most mundane aspects of our lives.

Living with Uncertainty

Together, the uncertainty principle and quantum tunneling reveal something profound about reality. The universe isn't the precise clockwork imagined by classical physics. Instead, it's built on a foundation of uncertainty and possibility.

This doesn't mean science is uncertain—quite the opposite. Quantum mechanics makes the most precise predictions of any scientific theory ever developed. We can calculate quantum effects with extraordinary accuracy. What's uncertain isn't our knowledge of the laws of physics but rather the inherent behavior of the particles themselves.

Modern technology increasingly embraces this quantum uncertainty. Quantum cryptography uses the uncertainty principle to create unbreakable codes. Quantum computers manipulate uncertain states to solve problems classical computers can't tackle. The future of technology may depend on learning to work with uncertainty rather than fighting against it.

The Edge of Knowledge

As we push the boundaries of science and technology, we keep running into these quantum limits. They remind us that the universe is stranger and more mysterious than our everyday experience suggests. But these limits haven't stopped progress—they've opened new doors.

Every time you use your phone, watch an LED screen or feel the sun's warmth, you're experiencing the effects of quantum uncertainty and tunneling. The same principles that limit our knowledge also enable our modern world.

When Particles Dance Together

We've seen how nature limits what we know about one particle. But something even stranger happens when we look at two particles together. They can share information instantly, no matter how far apart they are. Einstein called it "spooky." We call it quantum entanglement.

In the next chapter, we'll explore this mysterious connection between particles. This phenomenon challenged Einstein's view of reality and promised to revolutionize computing and communication. Get ready to discover why the quantum world isn't just uncertain—it's "spooky" too.

6

Entanglement - The Invisible Thread

"What happens to one particle can impact what we expect the second one to do, even if those particles are nowhere near each other." - David Kaiser.

Imagine having a pair of magical dice. You give one to your friend in Tokyo and keep one for yourself in New York. Every time you roll your die, your friend's die shows precisely the same number—instantly, with no communication between you. No radio signals, internet connection, or hidden transmitters— just two dice sharing an invisible bond across the planet. Impossible? Welcome to the world of quantum entanglement, perhaps the strangest feature of quantum physics and what Einstein famously called "spooky action at a distance."

The Twin Particle Mystery

In the quantum realm, particles can become "entangled" so that the fate of one is inextricably linked to the other, no matter how far apart they are. It's as if they share a secret connection that transcends space itself. When you measure one particle, you instantly know something about its distant partner.

Think of it as having identical twins who always wear opposite-colored socks without coordinating. If you see one twin wearing black socks in London, you instantly know their sibling in Sydney is wearing white socks. But even this analogy needs to capture the true strangeness of quantum entanglement. The socks were already black or white with the twins before you looked. In quantum entanglement, observing one particle determines the state of both particles at the exact moment.

Let's try another analogy. Imagine you have two coins sealed in separate envelopes. In the classical world, each coin is either a head or tail before you open the envelopes. But with quantum entanglement, it's as if the coins exist in a shared state of possibility until you open one envelope. At that instant, both coins snap into definite, correlated states, no matter how far apart the envelopes are.

Einstein's Spooky Action

When Einstein first encountered this idea, he was distraught. Here was the man who revolutionized our understanding of space and time with relativity, and he couldn't accept what quantum mechanics suggested about reality. His particular theory of relativity had shown that nothing, not even information, could travel faster than light. Yet here was quantum mechanics, suggesting that entangled particles could influence each other instantaneously across any distance.

Einstein believed the universe should make sense. Their immediate surroundings should only affect things like ripples spreading across a pond. The idea that measuring a particle in New York could instantly affect another particle on the moon seemed to violate everything he believed about how the universe should work.

In 1935, Einstein and his colleagues Boris Podolsky and Nathan Rosen published a paper that would become famous as the EPR paradox. They proposed a thought experiment that showed that quantum mechanics must be incomplete. Their argument was subtle but powerful: either quantum mechanics was wrong, or information could travel faster than light. Since Einstein had proven the latter was impossible, quantum mechanics had to be missing something.

Einstein suggested there must be "hidden variables"—some more profound layer of reality that would explain these impossible connections. In his view, quantum mechanics was like a weather forecast that could only predict probabilities. Just as a 70% chance of rain doesn't mean it's simultaneously raining and not raining, Einstein believed particles must have definite properties all along, even if we can't measure them all at once.

Bell's Brilliant Breakthrough

For nearly thirty years, the debate remained philosophical. How could you prove whether Einstein was right or wrong about something fundamental to reality? The answer would come from an unlikely source: an Irish physicist named John Bell, who worked in his spare time while developing particle accelerators at CERN.

In 1964, Bell found a way to settle the debate experimentally. He derived a mathematical inequality that could distinguish between Einstein's hidden variables and the predictions of quantum mechanics. If Einstein was right, certain correlations between entangled particles couldn't exceed a specific value. If quantum mechanics was correct, that limit could be violated.

The genius of Bell's work was that it turned a philosophical question into a testable prediction. It was like finding a way to scientifically test whether love exists by measuring its effects on the physical world. Bell showed that Einstein's common-sense view of reality made subtly different predictions from quantum mechanics.

To understand Bell's test, imagine our magical dice again. If the dice were using hidden variables—some pre-arranged pattern or synchronization—there would be limits to how perfectly they could correlate under different conditions. However, quantum mechanics predicted stronger correlations than should be possible with any hidden mechanism. It was as if the dice weren't just synchronized but shared a single existence despite being continents apart.

The scientific community realized Bell's work's profound implications. Here was a way to test whether quantum mechanics was complete, as its defenders claimed, or whether Einstein was right about hidden variables. If experiments could be designed carefully enough, nature would tell us which view was correct.

Proving Einstein Wrong: The Experiments

In 1972, physicist Alain Aspect and his team at the University of Paris set up an experiment that would shake the foundations of physics. The challenge was enormous: they needed to create pairs of entangled particles, send them in opposite directions, and measure their properties before any possible signal could pass between them. It was like trying to prove that twins could share thoughts while ensuring they had no way to communicate.

The experimental setup was ingenious. Using calcium atoms excited by lasers, they created pairs of entangled photons— particles of light that shared a quantum connection. These photons were sent racing away from each other through fiber optic cables. The team used unique crystals at precisely timed intervals to measure the photons' polarization—the direction in which their light waves oscillated.

If Einstein had been right about hidden variables, these measurements would have shown a limited correlation. Think of it like flipping two somehow synchronized coins before being separated. No matter how clever the synchronization, there's a mathematical limit to how often the coins can match up under different flipping conditions. Bell's inequality spelled out precisely what this limit should be.

But the photons violated this limit spectacularly. Their properties matched more often than any hidden variable theory could explain. It was as if the photons were consulting an invisible playbook transcending space and time.

Scientists, being scientists, were only partially satisfied. What if there were loopholes in the experiment? They identified several possibilities:

The Locality Loophole: What if some subtle signal could still pass between the particles? Teams started separating their detectors by increasing distances. In 2015, researchers at Delft University conducted tests with particles 1.3 kilometers apart. The spooky action persisted.

The Detection Loophole: What if the measuring devices were somehow selecting only the particles that would prove quantum mechanics right? Scientists developed more efficient detectors and statistical methods to ensure they weren't fooling themselves.

The Freedom-of-Choice Loophole: What if the universe some-how conspired to predict which measurements would be made? Researchers used random number generators based on distant stars to decide how to measure each particle, ensuring even the light from those stars couldn't have reached the other detector in time to influence it.

These loopholes were closed one by one. In 2018, an interna-tional team used light from distant quasars—ancient galaxies billions of light-years away—to determine their measurements. Even then, the quantum correlations held firm. Einstein's common-sense view of reality had been experimentally dis-proven not just once but hundreds of times under increasingly stringent conditions.

Creating Quantum Connections

How do we create entangled particles? It's not as mysterious as it might sound, though it does require precise control over quantum systems. Scientists have developed several methods, each with its advantages:

Spontaneous Parametric Down-Conversion (SPDC): This tongue-twister process uses unique crystals to split single photons into pairs of entangled photons with lower energy. Imagine breaking a playing card perfectly down the middle— the two halves will always match perfectly. The crystal does the same thing to light, creating photon pairs with a quantum connection.

Atomic Cascades: When excited to high energy states, some atoms emit pairs of photons as they return to their ground state. These photon pairs are naturally entangled, like twin children born simultaneously.

Quantum Dots: These tiny semiconductor structures can trap electrons and force them to emit entangled photons. It's like having a quantum light bulb that produces entangled particles on demand.

Recent breakthroughs have pushed the boundaries of what's possible. In 2023, Princeton physicists achieved something remarkable—entangled individual molecules for the first time. Using laser cooling to bring molecules nearly to absolute zero, they employed "optical tweezers" (focused laser beams that can trap and move tiny objects) to position molecules with incredible precision. Then, using carefully timed microwave pulses, they created quantum connections between these molecules.

Meanwhile, at CERN's Large Hadron Collider, scientists observed quantum entanglement between top quarks—the heaviest known elementary particles—and their antimatter counterparts. This showed that quantum effects don't just apply to tiny particles; they can manifest even at the highest energies we can create.

Creating quantum entanglement is just the first challenge—keeping it alive is where things get tricky. Imagine trying to preserve a soap bubble in a hurricane. That's what scientists face when working with quantum states. These delicate quantum connections are so fragile that even a stray air molecule or a whisper of heat can shatter them instantly. Scientists call this breakdown "decoherence," but you might think it is reality crashing in on quantum magic.

To protect these fragile quantum states, scientists have become masters of isolation. They build ultra-high vacuum chambers that make outer space look crowded, creating environments where particles can maintain their quantum dance undisturbed. They cool their experiments to temperatures so low they make the depths of space seem tropical—just a fraction of a degree above absolute zero. The atomic jiggling that could disrupt quantum states nearly stops at these temperatures.

When even that isn't enough, they employ sophisticated error correction techniques, like musical conductors keeping an orchestra in perfect harmony despite the occasional wrong note.

Some researchers work with exotic materials that naturally shelter quantum states from disruption, like building a fortress around these quantum connections. It's a constant battle against the tendency of the quantum world to lose its ethereal properties and collapse into the familiar classical behavior we see around us.

The Quantum Revolution in Security

Quantum entanglement isn't just a laboratory curiosity—it's becoming the foundation of a new type of unbreakable security. Traditional encryption relies on mathematical problems that are hard for computers to solve, but quantum encryption relies on the fundamental laws of physics themselves.

Banks have taken notice. In 2024, HSBC announced successful trials of quantum-safe technology for trading gold. JPMorgan Chase partnered with IBM to experiment with quantum algorithms for secure key exchange. These aren't just proof-of-concept experiments—they're the first steps toward a quantum-secure financial system.

China has taken the boldest step into this quantum future. In 2016, they launched Micius, the world's first quantum communication satellite. Named after an ancient Chinese philosopher, this satellite creates pairs of entangled photons in space and beams them to ground stations thousands of kilometers apart.

The achievement was remarkable. Think about what's happening: photons, created as quantum twins in space, rain down to Earth while maintaining their delicate quantum connection. 2017, this system enabled the first intercontinental quantum-encrypted video call between Beijing and Vienna. Any attempt to intercept the communication would have broken the quantum entanglement, immediately alerting both parties.

But this is just the beginning. China plans to launch a constellation of quantum satellites by 2025, creating a global network for unhackable communication. Other countries are racing to catch up. The European Union, the United States, and Japan have all announced major quantum communication initiatives.

Building the Quantum Internet

Imagine an internet where privacy is guaranteed not by software or mathematics but by the laws of physics itself. This is the promise of the quantum internet—a network that uses quantum entanglement to create unbreakable connections between users.

Unlike the classical internet, where information travels as bits (1s and 0s), the quantum internet will use quantum bits or "qubits." These qubits can be entangled with each other, creating fundamentally secure connections. Any attempt to eavesdrop would collapse the quantum state, making it impossible to intercept information without being detected.

The first quantum networks are already being built. In the Netherlands, researchers have connected four cities with a quantum network. China has built a 2,000-kilometer quantum communication link between Beijing and Shanghai. These are the first strands of what could become a global quantum web.

But building a quantum internet can be challenging. Quantum states are fragile and difficult to maintain over long distances. Scientists are developing "quantum repeaters"—devices that can extend quantum connections without breaking them. Think of them as relay stations for quantum information, preserving the delicate quantum dance of entangled particles.

Quantum Teleportation: Science Fiction Becomes Reality

When scientists first proposed quantum teleportation in 1993, it sounded like something from Star Trek. But today, it's a laboratory reality. Using quantum entanglement, scientists can instantly transfer the quantum state of one particle to another, regardless of the distance between them.

This isn't teleportation like in science fiction—you can't beam people or objects across space. What gets "teleported" is quantum information. But this ability to instantly transfer quantum states has profound implications for quantum computing and communication.

In 2020, Fermilab scientists achieved sustained, high-fidelity quantum teleportation over 44 kilometers of fiber optics. They successfully teleported qubits with an accuracy greater than 90%. Think about what this means: quantum information disappears from one location. It reappears in another faster than any classical signal could travel between the two points.

The Philosophical Implications

As we master quantum entanglement, we're forced to confront profound questions about the nature of reality. If particles can maintain instantaneous connections across any distance, what does this mean for our concepts of space and time? If measurement at one location can instantly affect a particle elsewhere, what happens to our ideas of cause and effect?

Some physicists, like Anton Zeilinger (who won the 2022 Nobel Prize for his work on entanglement), suggest that we must fundamentally revise our understanding of reality. The universe isn't made of separate objects that interact but of relationships and connections that define what we call reality.

The Observer Effect

We've seen how particles can stay connected across vast distances, maintaining their quantum dance of possibility until we measure them. But this leads us to an even stranger aspect of quantum physics: observation itself seems to change what we observe.

The next chapter will explore how looking at quantum systems changes their behavior. We'll see how consciousness and reality might be more deeply connected than we ever imagined and how this mysterious interaction shapes everything from laboratory experiments to the possibility of quantum computers.

Remember, every time you use a secure website or tap your phone to pay, you benefit from our understanding of quantum physics. The strange quantum dance of entangled particles isn't just a scientific curiosity—it's becoming the foundation of a technological revolution that will shape our future.

7

Transformation: Reality's Mirror

"The moon is demonstrably not there when nobody looks." That's what physicist David Mermin said about the strange world of quantum physics. It sounds absurd—surely the moon doesn't pop in and out of existence depending on who's watching—but this bizarre idea isn't just possible at the quantum level. It's fundamental to how reality works.

When Looking Changes Reality

Imagine you're watching a street performer. They're doing a fantastic dance routine, moving with incredible freedom and creativity. But when you try to record them with your phone, they freeze into a simple, static pose. This is similar to how quantum particles behave—observing them fundamentally changes their actions.

But quantum physics takes this idea to an extreme, making our street performer analogy seem tame. In the quantum world, things don't just behave differently when observed—they exist differently. It's not just that we're disturbing delicate quantum states with clumsy measurements. Reality itself seems to transform under our gaze.

Remember our discussion of Schrödinger's cat? That thought experiment wasn't just about cats in boxes—it was highlighting something profound about the nature of reality. Before we open the box (or make any quantum measurement), the system exists in all possible states simultaneously. It's not that the cat is either alive or dead. We don't know which—according to quantum mechanics, it's actually in a superposition of both states, existing in a blur of possibilities until the moment we look.

The Measurement Problem

This brings us to physics's most profound mysteries: the measurement problem. Why does the act of observation force quantum systems to "choose" one reality from all possible options? What counts as an observation? Does it require a conscious observer, or can any interaction with the environment cause this "collapse" of possibilities into certainty?

Imagine you're dealing cards from a deck. Every day, each card is a heart, diamond, club, or spade, even before you turn it face up. But if cards behaved like quantum particles, they would exist in a mixture of all possibilities until you looked at them. Even stranger, if you could peek at just the color of the card without seeing its complete identity, it would still maintain a superposition of all possibilities consistent with that partial information.

This isn't just philosophical musing—it has real, measurable consequences. Scientists can create particles in superposition states and watch what happens when they measure them. The results consistently show that reality behaves this way whether we like it or not. As physicist Richard Feynman famously said, "Nature isn't classical, dammit, and if you want to make a simulation of nature, you'd better make it quantum mechanical."

Making Sense of the Impossible

How do we make sense of this measurement mystery? Over the decades, physicists have proposed several interpretations with mind-bending implications.

The Copenhagen Interpretation, championed by Niels Bohr and his colleagues, says we should accept that quantum systems exist in multiple states until measured. At this point, they "collapse" into one definite state. It's like saying the rules of the quantum game are fundamentally different from our everyday experience, and we shouldn't try to force our classical intuitions onto the quantum world.

However, not everyone was satisfied with this "shut up and calculate" approach. Hugh Everett III proposed something even stranger: the Many-Worlds Interpretation. In this view, when we measure, it splits into multiple branches, each containing one of the possible outcomes. Every quantum choice creates new universes, endlessly branching like an infinite cosmic tree.

If this sounds like science fiction, you're not alone. The creators of "Avengers: Endgame" drew inspiration from these ideas. When the Avengers travel through time to collect the Infinity Stones, they're not changing their past—they're creating new branches of reality. Bruce Banner's insistence that "changing the past doesn't change the future" aligns perfectly with how quantum physicists think about parallel universes.

There's also the Pilot-Wave Theory, which suggests that particles always have definite positions and trajectories, guided by an invisible quantum wave that permeates all of space. This interpretation maintains determinism—the idea that everything has a definite cause and effect. Still, we can never directly observe without introducing a hidden layer of reality.

The Quantum Eraser: When the Present Changes the Past

When you think quantum mechanics couldn't get any stranger, enter the quantum eraser experiment. Imagine taking a photograph, but whether the picture shows a clear or blurry image depends on what you decide to do after you've already taken it. Sounds impossible? Welcome to the quantum world.

The experiment starts with our old friend, the double-slit setup. Scientists send particles through two slits, creating an interference pattern—unless they measure which slit each particle goes through, in which case the pattern disappears. But here's where it gets weird: they can "erase" this measurement information after the particles have already been detected.

What happens next defies common sense. When scientists

erase the "which-path" information, the interference pattern reappears—even though the particles were detected before the erasure decision was made. It's as if the present decision reaches backward in time to affect the particles' past behavior.

Think about what this means. You could perform the experiment, store the data on a hard drive, and fly that drive to another continent. Months later, when you finally decide whether to erase the which-path information, that decision affects how the particles behaved in the original experiment. As physicist John Wheeler put it, "The past has no existence except as it is recorded in the present."

Time's Arrow Bends

This quantum time-bending raises profound questions about causality—the relationship between cause and effect. In our everyday world, causes always precede effects. You can't win the lottery and then buy the ticket. You can't get a sunburn and then go to the beach. But quantum mechanics suggests that, at the microscopic level, time's arrow might not be so straight.

The "grandfather paradox" famously asks what would happen if you traveled back in time and prevented your grandparents from meeting. Similar paradoxes arise naturally in quantum mechanics. The quantum eraser experiment shows that future choices can affect past events, creating what appears to be a causal loop.

Reality's Hall of Mirrors

All these quantum peculiarities force us to confront some deeply uncomfortable questions about the nature of reality itself. If particles have definite properties once we measure them, what does this mean for the world when we're not looking at them? If observation can influence the past, what does this tell us about time? And perhaps most unsettling of all: what role does consciousness play in all of this?

Some physicists, including Eugene Wigner and John Wheeler, have suggested that consciousness might play a crucial role in collapsing quantum possibilities into definite realities. Others argue that any interaction with the environment—conscious or not—can cause this collapse. The debate touches on questions that blur the line between physics and philosophy: What is consciousness? What makes a measurement a measurement? Where exactly does the quantum world end and our classical world begin?

Think about a tree falling in a forest with no one around. The old philosophical question asks if it makes a sound. Quantum mechanics raises an even stranger possibility: does the tree occupy a definite position before someone looks at it? Does it exist in a superposition of all possible states until observed? These aren't just philosophical musings—they're questions about the fundamental nature of reality itself.

The Observer Effect in Your Life

While these ideas might seem abstract, they have real implications for modern technology. Quantum cryptography uses the observer effect to create unbreakable codes—any attempt to intercept the message necessarily changes it, alerting the intended recipients to eavesdropping. Quantum computers manipulate particles in superposition states to perform calculations impossible for classical computers.

Even some biological systems might exploit quantum effects. Scientists have found evidence that birds use quantum entanglement in their magnetic navigation systems, and photosynthesis may rely on quantum coherence to achieve its remarkable efficiency. Nature has been using quantum mechanics long before we understood it.

The Ultimate Mystery

The most profound implication of quantum measurement is what it tells us about our place in the universe. We're not just passive observers watching reality unfold—we're active participants in determining what becomes real. Every measurement we make, every observation we choose to make or not make, helps shape the reality we experience.

As some New Age interpretations suggest, this doesn't mean we create reality with our minds. Instead, it means the classical world of definite properties and transparent cause-and-effect relationships emerges from a deeper quantum reality through a process we're only beginning to understand.

Quantum Technology in Action

Now that we've explored how quantum systems react to observation—how reality itself seems to transform under our gaze—it's time to see how we're putting this knowledge to work. In the next chapter, we'll explore the quantum technologies already changing our world, from quantum computers that harness superposition to perform impossible calculations to quantum sensors that can detect single atoms.

Remember, if these ideas seem strange, you're in good company. As physicist Richard Feynman once said, "I can safely say that nobody understands quantum mechanics." That may be the point. Quantum mechanics doesn't just describe the world—it reveals that it is far stranger and more mysterious than we ever imagined.

The quantum technologies we're building today aren't just clever applications of well-understood principles—they're explorations of the deepest mysteries of reality itself. Every quantum computer, every quantum sensor, and every quantum encryption system is also a philosophical experiment, probing the boundary between the observer and the observed, between the possible and the actual, and between the quantum and classical worlds.

As we move into this quantum future, we're not just developing new technologies—we're creating new ways of thinking about reality. And perhaps that's the most transformative aspect of quantum physics: it shows us that the universe is not just stranger than we imagine but stranger than we can imagine.

III

Quantum Technology and the Future

8

Innovation - Tomorrow's Technology

"Nature isn't classical, dammit, and if you want to make a simulation of nature, you'd better make it quantum mechanical." Those are the words of physicist Richard Feynman, who inspired the creation of quantum computers. Today, his vision is becoming a reality—changing the world faster than you think.

From Bits to Qubits: A New Kind of Computing

Imagine trying to read a book in a pitch-dark room with just a flashlight. You can only see one word at a time, checking each word sequentially until you find what you're looking for. This is essentially how classical computers work with bits—examining information piece by piece, each bit either a 0 or 1, like a light switch on or off.

Imagine if you could flood the entire room with light, reading every word simultaneously. This is the promise of quantum computing. Instead of bits, quantum computers use qubits—

quantum bits that can exist in multiple states simultaneously, like a dimmer switch on and off simultaneously.

But this analogy, while helpful, barely scratches the surface of what makes quantum computing revolutionary. A better way to think about it might be like this: Imagine trying to find your way out of an enormous maze. A classical computer would try one path at a time, backtracking when it hits dead ends. A quantum computer, through the strange properties of superposition, effectively explores all possible paths simultaneously. It's not just faster—it's fundamentally different.

The power of qubits grows exponentially. While 100 classical bits can only be in one of 2^{100} possible combinations simultaneously, 100 qubits can simultaneously exist in all 2^{100} states. That's more states than there are atoms in the visible universe. This isn't just an improvement in computing power—it's a new paradigm that defies our classical intuitions.

The Dance of Quantum Information

Creating and controlling qubits is like choreographing an impossibly delicate dance. Scientists use various physical systems to create qubits: superconducting circuits cooled to near absolute zero, individual atoms trapped by lasers, or electrons whose quantum spin can point in multiple directions simultaneously. Each approach has its challenges and advantages, but they all share one critical feature: they harness the weird properties of quantum mechanics to process information in ways classical computers never could.

Think about shuffling a deck of cards. A classical computer would have to look at each possible arrangement one at a time. But a quantum computer could consider all possible shuffles at once. With just 52 cards, there are more possible arrangements than stars in the visible universe. A quantum computer could process all these possibilities simultaneously.

This massive parallelism comes from the quantum properties discussed in previous chapters. Superposition allows each qubit to be in multiple states at once. Entanglement lets qubits share information in ways that have no classical equivalent. Even quantum tunneling, which we explored earlier, plays a role in some quantum computing architectures.

The Challenge of Quantum Coherence

But there's a catch—and it's a big one. Quantum states are incredibly fragile. The slightest interaction with the environment can cause "decoherence," destroying the delicate quantum properties that give qubits their power. It's like trying to balance a pencil on its tip while the Earth trembles from millions of tiny earthquakes.

To maintain quantum coherence, scientists go to extraordinary lengths. They cool their quantum computers to temperatures colder than deep space, just fractions of a degree above absolute zero. They use sophisticated error correction techniques to protect quantum information. They develop new materials and methods to shield their qubits from noise and interference.

Despite these challenges, quantum computers are already

achieving things that would be impossible with classical computers. They're simulating complex molecular interactions for drug discovery, optimizing financial portfolios with unprecedented sophistication, and even helping us understand the quantum nature of reality itself.

Solving the Impossible: What Quantum Computers Do Best

In 2019, Google announced something extraordinary. Their quantum computer had solved a problem in 200 seconds that would have taken the world's most powerful classical supercomputer 10,000 years to crack. They called it "quantum supremacy," the moment when a quantum computer accomplishes something effectively impossible for classical computers.

But what kinds of problems are quantum computers uniquely suited to solve? They excel at tasks that classical computers find nightmarishly tricky. Imagine trying to factor a number like 2,048 into its prime components. Is it simple? Now try it with a number that's hundreds of digits long. Classical computers struggle with this task so much that we use this difficulty to encrypt much of our digital communication. Quantum computers, however, could crack such codes in minutes.

Yet code-breaking is just the beginning. In pharmaceutical laboratories, quantum computers are revolutionizing drug discovery. Traditional computers struggle to simulate complex molecular interactions—there are too many quantum effects to account for. But quantum computers speak the language of quantum mechanics natively. They can model how drugs interact with target molecules, potentially reducing the time and cost of developing new medicines from decades to years.

Nature's Quantum Calculator

Climate scientists are particularly excited about quantum computing's potential. Our current climate models, running on classical supercomputers, can only approximate molecular systems' incredibly complex quantum interactions. Quantum computers could simulate these systems, helping us understand and combat climate change more effectively.

They're also transforming material science. Want to design a better battery for electric cars, a more efficient solar panel, or a superconductor that works at room temperature? These challenges require understanding how electrons interact in complex materials—a problem tailor-made for quantum computers.

The AI Revolution Gets Quantum

Most intriguingly, quantum computers could supercharge artificial intelligence. Classical AI already seems impressive, but the sequential nature of traditional computing limits it. Quantum computers could process vast neural networks simultaneously, potentially leading to AI systems with capabilities we can barely imagine.

Think about how your phone's AI assistant recognizes your voice. It compares your speech patterns against countless possibilities until it finds a match. A quantum computer could theoretically check all possibilities simultaneously, making voice recognition instantaneous and perfect. Similar principles could revolutionize everything from image recognition to language translation.

The Quantum Internet: A New Kind of Network

While we're building ever-more-powerful quantum computers, scientists are also working on something equally revolutionary: the quantum internet. Unlike our current internet, which sends information as classical bits, the quantum internet will transmit quantum states, using entangled particles to create unhackable communication channels.

Imagine sending a message that's not just encrypted but protected by the laws of physics themselves. Any attempt to intercept the message would collapse its quantum state, immediately alerting both sender and receiver to the intrusion. China has already launched a quantum communication satellite demonstrating this technology, and other countries are racing to build their quantum networks.

But the quantum internet isn't just about secure communication. It could also connect quantum computers, creating a network of unprecedented computational power. Think of it as a quantum cloud, where the strange properties of quantum mechanics are available on demand.

Quantum Space: The Final Frontier

Space exploration is another frontier where quantum technology promises breakthroughs. Quantum sensors could detect gravitational waves with unprecedented precision, helping us understand the cosmos in new ways. Quantum navigation systems could guide spacecraft without relying on signals from Earth. Some scientists even speculate about quantum propulsion systems that could make interstellar travel possible.

NASA and other space agencies already use quantum technology to improve their measurements and communications. The exact quantum properties that make secure communication possible on Earth could one day help us stay in touch with colonies on Mars or probes in deep space.

The Near Future: What's Coming Next

We're standing at the threshold of a quantum revolution. Within the next decade, we'll likely see:

- Quantum computers solve problems in chemistry and materials science that are impossible today
- Quantum sensors detecting diseases at the molecular level before symptoms appear
- Quantum communication networks spanning continents
- Quantum financial systems that can spot market patterns invisible to classical computers

But the most exciting applications are the ones we haven't even imagined yet. When classical computers were first invented, no one predicted smartphones or social media. The quantum revolution may bring equally unexpected innovations.

A Note of Caution

Of course, with great power comes great responsibility. Quantum computers could eventually crack many encryption systems that protect our digital lives. That's why researchers are already developing "quantum-safe" encryption methods that even quantum computers can't break.

There are also concerns about quantum computing widening the technological divide between nations. Countries that master quantum technology first could gain significant advantages in everything from medicine to military applications.

The Quantum Body

While we've focused on quantum technology in this chapter, something equally fascinating awaits us: quantum biology. Your body might use quantum effects in ways we're only beginning to understand. From the efficiency of photosynthesis to the mystery of bird navigation, from the mechanics of smell to the possibility of quantum effects in consciousness, nature may have been doing quantum computing long before we invented it.

In the next chapter, we'll explore how quantum mechanics isn't just about technology—it's about life itself. We'll discover how plants use quantum coherence to turn sunlight into energy with near-perfect efficiency, how birds might use quantum entanglement to navigate using Earth's magnetic field, and how the very mechanisms of evolution might depend on quantum effects.

Remember, when Feynman said nature isn't classical, he wasn't just talking about laboratory physics experiments. He was talking about everything—including us. The quantum revolution is changing our technology and our understanding of life itself.

9

Medicine - Nature's Secret

"Nature uses only the longest threads to weave her patterns so that each small piece of her fabric reveals the organization of the entire tapestry." Richard Feynman wasn't just being poetic when he said these words—he was hinting at something profound about how the universe works, from the most minor quantum effects to the most prominent living systems.

We tend to think of quantum physics as something that only happens in laboratories, with sophisticated equipment cooled to near absolute zero. But nature has been doing quantum physics for billions of years. It's doing it right now, in every cell of your body, every leaf of every plant, and every bird's brain soaring overhead. While our most advanced quantum computers require temperatures colder than deep space to maintain quantum states for mere seconds, living things manipulate quantum systems at room temperature with remarkable precision.

The Quantum Compass in Birds' Eyes

Millions of birds embark on one of nature's most remarkable journeys every autumn. Take the Arctic tern, for instance. This small seabird navigates yearly from the Arctic to the Antarctic—a round trip of about 44,000 miles. It's like flying around the Earth's equator one and a half times, often through utterly unfamiliar territory, yet these birds rarely get lost. Even more impressive, they can navigate on cloudy days when they can't see the sun or stars and in conditions where familiar landmarks are nowhere to be seen.

For decades, this incredible feat of navigation puzzled scientists. They knew birds could sense Earth's magnetic field—a discovery that seemed like magic when first proposed. But the mechanism remained mysterious. The Earth's magnetic field is incredibly weak—about a hundred times weaker than a refrigerator magnet. How could birds detect something so subtle? And how could they use this information to determine not just direction but their position on the globe?

The answer lies in one of quantum mechanics' strangest properties: entanglement. Deep in the birds' eyes sits a protein called cryptochrome. When light hits this protein, it triggers a quantum reaction that creates a pair of entangled electrons. These electrons are exquisitely sensitive to magnetic fields thanks to a quantum property called spin. The interaction between these entangled electrons and Earth's magnetic field creates distinct quantum states that the bird's brain can interpret as directional information.

This quantum compass is so sophisticated that it shames our best navigation technology. While GPS needs at least four satellites to determine your position, birds carry a quantum positioning system in their eyes that works anywhere on Earth. They're using quantum entanglement—the same property Einstein found so troubling he called it "spooky action at a distance"—every time they migrate.

The discovery of this quantum navigation system has profound implications. First, it shows that quantum effects aren't just microscopic curiosities—they can influence behavior in large organisms. Second, it demonstrates that nature has solved one of our biggest challenges in quantum technology: maintaining quantum states at room temperature. While we struggle to keep quantum states coherent for microseconds in carefully controlled laboratory conditions, birds maintain and use quantum systems in their bodies' messy, warm environments.

Scientists are studying bird navigation to understand how nature achieves this quantum stability. The insights could help us design better quantum computers, develop more precise navigation systems, and create new medical imaging technologies. Some researchers speculate that understanding how birds process quantum information in their brains might give us clues about consciousness.

But the quantum mysteries of nature don't end with bird navigation. Even more remarkable quantum effects are at work in every green leaf and blade of grass around us, orchestrating one of life's most fundamental processes: photosynthesis.

The Quantum Dance of Photosynthesis

Every second, plants worldwide perform a miracle that powers almost all life on Earth: photosynthesis. When sunlight strikes a leaf, something remarkable happens. The light's energy must travel through a complex network of molecules to reach the reaction center, where it can be converted into chemical energy. According to classical physics, this should be inefficient, with much energy lost along the way. Yet plants achieve near-perfect efficiency in transferring this energy.

The secret? Quantum coherence. When light energy enters a leaf, instead of taking a single path through the molecular maze, it takes all possible paths simultaneously, just like a quantum particle passing through multiple slits. This quantum superposition allows the energy to find the most efficient route instantly. It's as if each photon of sunlight explores every possible path through the leaf at once, automatically choosing the best one.

Scientists discovered this quantum behavior by shooting laser pulses at photosynthetic bacteria and watching the energy move through their light-harvesting complexes. They saw something astonishing: wave-like patterns that could only be explained by quantum mechanics. The energy wasn't traveling like a classical particle bouncing through a pinball machine—it was moving like a quantum wave, spreading out and interfering with itself to find the optimal path.

What makes this even more remarkable is that it happens at room temperature, in a living cell's wet, chaotic environment.

The quantum states in our best laboratories fall apart in a fraction of a second, yet plants maintain quantum coherence long enough to capture sunlight efficiently. They've evolved what engineers would call a "quantum heat engine," operating with an efficiency that human technology can't match.

Nature's Quantum Technology

The implications of these discoveries stretch far beyond biology. Engineers are now studying photosynthesis to design better solar cells. If we could replicate nature's quantum efficiency, we could revolutionize renewable energy. Some scientists are even developing "artificial leaves" that use quantum effects to split water into hydrogen and oxygen, potentially creating clean fuel from sunlight.

But nature's quantum engineering continues beyond photosynthesis and bird navigation. Recent research suggests quantum effects work in our sense of smell. The traditional theory says smell works by molecules fitting into receptors like keys into locks. However, we must explain how we can distinguish between molecules with identical shapes but different isotopes. The answer might lie in quantum tunneling—the same effect that makes transistors work in your computer might help you distinguish the scent of a rose from a lily.

Even enzymes, the proteins that catalyze chemical reactions in our bodies, might use quantum tunneling to speed up their work. When an enzyme needs to transfer a hydrogen atom from one molecule to another, sometimes the atom doesn't go over the energy barrier—it tunnels through it, just like the quantum particles we discussed earlier.

The Future of Quantum Medicine

While plants and birds have been using quantum effects for millions of years, we're just beginning to harness quantum mechanics for human health. A new field called quantum medicine is emerging, promising to revolutionize how we diagnose and treat diseases.

Consider quantum imaging technology. Traditional MRI machines already use quantum effects—the spin of atomic nuclei—to create images of our bodies. However, new quantum sensors could detect individual molecules that are out of place, potentially spotting diseases like cancer years before they become visible through conventional methods. It's like giving doctors quantum glasses that let them see the earliest signs of illness at the molecular level.

Drug discovery is another field being transformed by quantum science. Understanding how molecules interact is fundamentally a quantum mechanical problem—electrons jumping between energy levels and atoms sharing quantum states to form chemical bonds. Classical computers struggle to simulate these interactions accurately, so developing new drugs traditionally takes years and billions of dollars. Quantum computers, however, speak the language of quantum mechanics natively. They could simulate molecular interactions with unprecedented accuracy, potentially reducing drug development time from decades to months.

Some scientists are even exploring how quantum effects might help target cancer cells more precisely. Traditional radiation therapy can damage healthy tissue along with cancer cells. However, quantum-controlled radiation could theoretically target diseased cells with atomic precision, minimizing side effects and maximizing treatment effectiveness.

The Quantum Brain

The most intriguing frontier in quantum biology is the possibility that quantum effects play a role in how our brains work. The human brain contains roughly 86 billion neurons connected to thousands of others, creating a network of unimaginable complexity. Could quantum mechanics help explain how this intricate system gives rise to consciousness?

Nobel laureate Roger Penrose and anesthesiologist Stuart Hameroff have proposed that quantum processes in structures called microtubules—tiny tubes that help maintain cellular structure—might be crucial to consciousness. While their specific theory remains controversial, the idea that quantum effects might influence brain function is gaining scientific attention.

Recent research has found evidence of quantum coherence in the brain's molecular machinery. The exact quantum effects that help plants harvest sunlight efficiently might help our neurons process information. Some scientists speculate that quantum entanglement between molecules in different neurons could help explain the brain's remarkable information-processing abilities.

Nature's Quantum Legacy

As we've seen throughout this chapter, nature didn't just stumble upon quantum mechanics—it mastered it. From the quantum compass in birds' eyes to the quantum coherence in photosynthesis, from the possible quantum effects in our sense of smell to the mysteries of consciousness, living things have been exploiting quantum phenomena for billions of years.

This realization is transforming how we think about biology and medicine. We're learning that life isn't just subject to quantum effects—it actively harnesses them. Each discovery of quantum processes in nature provides new insights that could help us develop better technologies and treatments.

The Quantum Universe

We've seen how quantum physics shapes life on Earth, from the tiniest cellular processes to the possibility of quantum effects in consciousness. But the quantum story is much more significant. In the next chapter, we'll see how quantum mechanics shapes life and the entire universe—from the nuclear furnaces of stars to the vast cosmic web of galaxies.

Remember, every breath you take, every beat of your heart, every thought in your mind might depend on quantum effects that we're only beginning to understand. The quantum world isn't just about subatomic particles in physics labs—it's about us, life itself, and our place in this remarkably quantum universe.

10

Expansion - The Quantum Universe

"What we have called matter is energy, whose vibration has been so lowered as to be perceptible to the senses. There is no matter." When Einstein made this statement, he was hinting at something profound: the universe, at its deepest level, isn't made of tiny billiard balls we call particles. Instead, reality emerges from something far stranger—a cosmic quantum dance of fields and energy that stretches across space and time.

The Symphony of Fields

Imagine sitting by a still pond on a quiet morning. Drop a pebble, and waves ripple across the water's surface, creating circular patterns that spread outward. Drop two pebbles, and their ripples interfere, creating complex patterns of peaks and troughs. Now imagine the entire universe filled not with water but invisible fields permeating every point in space. These aren't like magnetic or electric fields you might remember from school—they are quantum fields and the foundation of everything we see around us.

According to quantum field theory, our universe is more like a vast collection of overlapping symphonies than a container full of particles. Just as a symphony hall fills with overlapping sound waves from different instruments, space fills with overlapping quantum fields. Each type of particle we observe—electrons, photons, quarks—corresponds to its universal field that extends throughout space and time.

Particles are localized vibrations in these fields, like notes played on cosmic instruments. When we observe an electron, we're not seeing a tiny sphere of matter—we're detecting a particular kind of vibration in the electron field, much like hearing a specific note in a symphony. When we see light, we're not observing little particles streaming through space—we're detecting oscillations in the electromagnetic field.

These quantum fields interact and influence each other in precise mathematical ways. When an electron seems to emit or absorb a photon, the electron field and electromagnetic field exchange energy, creating what we perceive as particle interactions. It's like two musical instruments playing together, creating harmonies and resonances that we interpret as physical events.

The mathematics of quantum field theory makes precise predictions about how these fields interact. It successfully explains phenomena that baffled earlier physicists, like how particles can pop into existence from apparent space or how forces can act across distances. The theory is so accurate that some of its predictions have been confirmed to a precision of one part in a trillion—like measuring the distance between New York and Los Angeles to the width of a human hair.

This view of reality revolutionizes our understanding of matter itself. The solid objects we see around us—tables, chairs, even our bodies—are not fundamentally different from the forces that act between them. Both matter and forces emerge from the same underlying quantum fields. The distinction between substance and energy, particle and wave, matter and force dissolves at this most profound level of reality.

The discovery of the Higgs boson in 2012 dramatically confirmed quantum field theory. The Higgs field was predicted to exist because the mathematics of quantum fields required it—without it, other particles couldn't have mass. When the Large Hadron Collider detected the predicted vibration in this field (the Higgs boson), it validated our understanding of how quantum fields create the properties of matter we observe.

The Quantum Nature of Nothing

What's in space? Our common sense tells us "nothing," but quantum field theory reveals something more fascinating. Even in the emptiest region of space, quantum fields are never truly still. They fluctuate constantly, like a calm ocean that's never wholly flat but always has tiny ripples moving across its surface.

These quantum fluctuations have remarkable consequences. According to the uncertainty principle we discussed earlier, we can never know the strength of a field and how quickly it

EXPANSION - THE QUANTUM UNIVERSE

changes simultaneously. This means that even in a vacuum, fields must maintain minimal activity—a sort of quantum jitter that can never be eliminated.

This isn't just theoretical speculation. These vacuum fluctuations have measurable effects in the real world. Consider the Casimir effect, first predicted by Dutch physicist Hendrik Casimir in 1948. Place two metal plates extremely close together in a vacuum, and they'll experience a tiny attractive force. Why? The quantum fluctuations between the plates are restricted to specific wavelengths, while all wavelengths are possible outside the plates. This difference in quantum activity creates a small but measurable pressure pushing the plates together.

Even more remarkably, these vacuum fluctuations can momentarily create pairs of particles and antiparticles that pop into existence and then quickly annihilate each other. Physicists call these "virtual particles," while they exist too briefly to be directly observed, their effects can be measured. They contribute to the properties of atoms, affect the strength of electromagnetic forces, and might even play a role in the behavior of black holes through Hawking radiation.

Think about what this means: the vacuum isn't empty at all. It's a seething quantum foam of virtual particles appearing and disappearing, fields fluctuating between positive and negative values, and energy borrowing itself from nothing for the briefest moments allowed by the uncertainty principle. Every cubic centimeter of "empty" space contains more activity than we could imagine.

Dark Mysteries and Quantum Solutions

This quantum view of space might help us understand two of the biggest mysteries in modern cosmology: dark matter and dark energy. These invisible components make up 95% of the universe's content, yet we still don't know what they are.

Dark matter reveals itself through its gravitational effects on galaxies and galaxy clusters. It forms an invisible scaffold that holds cosmic structures together. While we can't see it directly, some theorists suggest it might be made of particles that interact through yet-undiscovered quantum fields. These hypothetical particles rarely interact with ordinary matter except through gravity, making them extremely difficult to detect.

Dark energy is even more mysterious. It is a property of space itself, pushing the universe to expand at an accelerating rate. Some physicists speculate that dark energy might be related to vacuum energy—the energy of quantum fluctuations in space. The challenge is that when we calculate how much vacuum energy there should be using quantum field theory, we get a number that's way too large. This "cosmological constant problem" remains one of physics's most giant unsolved puzzles.

When Reality Multiplies

The quantum nature of the universe raises an even more mind-bending possibility: our universe might be just one of many. Remember the Many-Worlds Interpretation of Quantum Mechanics we discussed earlier? It suggests that every quantum event creates new branches of reality. But quantum field theory takes this idea even further, hinting at the possibility of multiple universes emerging from the quantum vacuum itself.

Think about those virtual particles we discussed, popping in and out of existence in space. Some theories suggest that similar quantum fluctuations, operating at a cosmic scale, might give birth to entire universes. Each would bubble up from the quantum vacuum like a bubble forming in boiling water, expanding to create its own space, time, and physical laws.

String theory, an ambitious attempt to unite quantum mechanics with gravity, makes this multiverse idea even more intriguing. It suggests that we think of particles as tiny vibrating strings of energy, like infinitesimal violin strings playing cosmic melodies. These strings would vibrate in extra dimensions we can't see, creating all the particles and forces we observe.

Just as a guitar string can vibrate in different patterns to produce different notes, these cosmic strings could vibrate in countless ways, potentially creating various particles and forces in other universes. With its particular set of physical laws and constants, our universe might be just one possible "song" these strings can play.

If you've watched "Doctor Strange in the Multiverse of Madness," you've seen a Hollywood version of these ideas. While the movie takes creative liberties, the core concept—that reality might be far more expansive than our single universe—emerges naturally from our best physical theories.

The Quantum Tapestry

A remarkable picture emerges as we zoom out from quantum fields to multiple universes. The exact quantum principles we discovered by studying tiny particles shape the most significant structures in the cosmos. The uncertainty principle affecting electron positions helps create new universes, and the vacuum fluctuations influencing atomic properties drive cosmic expansion.

Every atom in your body exists as a set of vibrations in quantum fields that extend throughout space and time. These fields interact according to precise mathematical rules, creating the intricate dance we call reality. The space around you teems with virtual particles and quantum fluctuations. And our entire universe might be one bubble in an infinite quantum foam of possible realities.

This view of the cosmos raises profound questions. If our universe is one of many, what does this mean for the laws of physics? Are they universal constants or just local bylaws? What happens to all those other possibilities if reality branches every time a quantum choice is made? And perhaps most fundamentally, why does our universe have the properties that allow for our existence?

The Nature of Reality

We've traveled from the tiniest quantum particles to the most significant structures in the universe. We've seen how quantum physics shapes everything—from the atoms in your body to the stars in the sky, from the vacuum of space to the possibility of multiple universes. Now, it's time to step back and ask what this all means for our understanding of reality.

In the final chapter, we'll explore the philosophical implications of quantum physics. We'll confront questions about the nature of reality, the role of consciousness in the quantum world, and what quantum mechanics tells us about our place in this vast, strange, and beautiful cosmos.

Remember, when Einstein said there is no matter, he wasn't being metaphorical. The solid reality we experience emerges from an underlying quantum realm that challenges our deepest intuitions about what is real. As we've seen throughout this book, the universe is not only stranger than we imagine—it's stranger than we can imagine. But through quantum physics, we're beginning to glimpse the true nature of this cosmic quantum dance.

11

Conclusion: Your Quantum Journey

When we began this journey through quantum physics, we started with a simple idea: that the universe, at its deepest level, behaves in ways our everyday experience could never prepare us for. Having explored everything from the double-slit experiment to quantum computers, from entangled particles to the possibility of multiple universes, we can see just how profound this insight truly is.

Quantum physics isn't just about tiny particles or abstract mathematics—it's about the fundamental nature of reality itself. Every photon in a sunbeam, an electron in a lightning bolt, and an atom in your body follows quantum rules. The solid world we experience emerges from a deeper quantum reality where particles can exist in multiple places simultaneously, where space teems with virtual particles, and where the mere act of observation can change the outcome of an experiment.

But most remarkably, we've seen how nature harnesses quantum effects with extraordinary precision. Birds navigate using quantum entanglement in their eyes. Plants achieve near-perfect efficiency in photosynthesis through quantum coherence. Even our consciousness might depend on quantum processes in our brains. While we struggle to maintain quantum states for microseconds in our most advanced laboratories, living things have been doing quantum physics for billions of years.

The technologies we build based on these quantum principles are already transforming our world. Quantum computers are tackling problems that classical computers could never solve. Quantum cryptography promises unbreakable codes protected by the laws of physics themselves. Quantum sensors might soon detect diseases at the molecular level, revolutionizing medicine.

We've broken down these complex ideas into understandable pieces through the S.P.A.C.E.T.I.M.E. framework. You now grasp concepts that once seemed impossible to comprehend. You understand why Einstein called entanglement "spooky action at a distance," Schrödinger put his famous cat in a box, and Feynman said, "Nobody understands quantum mechanics."

But this isn't the end of your quantum journey—it's just the beginning. Scientists are making discoveries about how quantum mechanics shapes our world daily. As you go forward, keep asking questions. Stay curious about the quantum nature of reality. Remember that every technology you use, from your smartphone to your GPS, works because someone dared to explore the quantum realm.

Please leave a review if you've found this book helpful in understanding the quantum world. Your feedback helps other curious minds navigate these fascinating ideas.

The universe is far stranger and more wonderful than we ever imagined. But we're beginning to understand its most profound mysteries through quantum physics. Keep exploring, keep learning, and keep marveling at the quantum dance that underlies all of reality.

Glossary: Understanding the Language of Quantum Physics

Amplitude: In quantum mechanics, a complex number that determines the probability of finding a particle in a particular state. The square of the amplitude's magnitude gives the likelihood of measuring that state.

Antiparticle: A particle with the same mass but opposite properties (like charge) as its regular counterpart. When particles meet their antiparticles, they annihilate each other, converting their mass into pure energy.

Bell's Theorem: A mathematical proof by John Bell showing that quantum mechanics predicts stronger correlations between particles than any theory based on local hidden variables could explain. It demonstrates that quantum entanglement is genuine and can't be explained by classical physics.

Bohr Model: An early model of the atom proposed by Niels Bohr, suggesting that electrons orbit the nucleus in fixed energy levels, like planets around the sun. While superseded by quantum mechanics, it helped establish quantized energy states.

Bose-Einstein Condensate: A state of matter formed when

particles called bosons are cooled to temperatures close to absolute zero. In this state, multiple particles occupy the same quantum state, behaving as a single quantum entity.

Copenhagen Interpretation: This is the traditional interpretation of quantum mechanics, developed primarily by Niels Bohr and Werner Heisenberg. It states that quantum systems exist in all possible states simultaneously until measured. At this point, they "collapse" into a definite state.

Coherence: The property that allows quantum systems to exist in multiple states simultaneously. When a system is coherent, its quantum waves remain in step and can interfere with each other, like ripples on a pond.

Dark Energy: A mysterious form of energy that appears to permeate all of space and drives the universe's accelerating expansion. It might be related to quantum vacuum energy.

Dark Matter: An invisible form of matter that interacts through gravity but not through electromagnetic forces. It might be composed of yet-undiscovered quantum particles.

Decoherence: The process by which quantum systems lose their quantum properties through interaction with their environment. It explains why we don't see quantum effects in everyday objects.

Eigenstate: A quantum state with a definite value for a particular observable property. When measured, a system in an eigenstate will always yield the same result for that property.

Entanglement: A quantum connection between particles where measuring one instantly determines the state of the other, no matter how far apart they are. Einstein called this "spooky action at a distance."

Fermion: A type of particle following the Pauli exclusion principle, meaning no two fermions can occupy the same quantum state. Electrons, protons, and neutrons are all fermions.

Field: In quantum field theory, a field exists at every point in space and can oscillate like a wave. Particles are excitations in these fields, like waves on an ocean.

Gauge Theory: A type of quantum field theory that describes force-carrying particles (like photons) arising from natural symmetries.

Heisenberg Uncertainty Principle: A fundamental limit of nature stating that we cannot simultaneously know both a particle's position and momentum with perfect precision. The more accurately we measure one, the less we know about the other.

Interference: The phenomenon where quantum waves combine, reinforcing or canceling each other. This creates patterns like those seen in the double-slit experiment, demonstrating the wave nature of quantum objects.

Many-Worlds Interpretation: A theory proposing that quantum measurements don't collapse the wave function but instead split reality into multiple branches, each containing one possible outcome. According to this interpretation, all possible quantum outcomes occur in different parallel universes.

Measurement Problem: The fundamental question in quantum mechanics is why and how quantum systems appear to change their behavior when measured. This problem lies at the heart of debates about quantum interpretation.

Observer Effect: The principle that measuring a quantum system inevitably affects it. This is different from the uncertainty principle—it's about the physical interaction required to make a measurement.

Orbital: In quantum mechanics, a region of space where an electron is likely to be found around an atom. Unlike the circular orbits of the Bohr model, orbitals are probability clouds with complex three-dimensional shapes.

Pauli Exclusion Principle: A fundamental rule stating that no identical fermions (like electrons) can simultaneously occupy the same quantum state. This principle explains the atomic structure and why the matter is solid.

Photon: A quantum of electromagnetic radiation—the particle of light. Photons have no mass, always travel at the speed of light, and demonstrate wave and particle properties.

Pilot Wave Theory: An alternative interpretation of quantum mechanics proposing that particles have definite positions and are guided by an actual wave. This theory maintains determinism but requires instantaneous action at a distance.

Planck Constant: A fundamental physical constant (h) that defines the scale at which quantum effects become important. It relates a particle's energy to its wavelength and is crucial in many quantum mechanical equations.

Planck Length: The minor meaningful length scale in physics (about ten^{-35} meters), where quantum effects and gravity become equally important. Below this scale, our current physics theories break down.

Probability Amplitude: A complex number used in quantum mechanics whose square gives the probability of measuring a particular outcome. Unlike classical probabilities, amplitudes can be harmful or complex numbers.

Probability Cloud: A region of space where a quantum particle is likely to be found when measured. The denser the cloud at a particular point, the more likely the particle will be detected.

Quantum Chromodynamics (QCD): The theory describing the strong nuclear force that binds quarks together to form protons, neutrons, and other particles. It explains why quarks are never found alone but always in combinations.

Quantum Cryptography: A secure communication method that uses quantum mechanics principles, particularly entanglement and the observer effect, to detect eavesdropping attempts. Unlike classical encryption, its security is guaranteed by the laws of physics.

Quantum Electrodynamics (QED): The quantum theory of the electromagnetic force, describing how light interacts with matter. It's one of science's most precisely tested theories, accurately predicting electromagnetic effects to within one part in a trillion.

Quantum Entanglement: When two or more particles become correlated so that the quantum state of each cannot be described independently. Measuring one particle instantly determines the state of its entangled partners, regardless of distance.

Quantum Field: A field that permeates all of space and can exhibit quantum mechanical properties. Different types of particles are understood as excitations in their corresponding quantum fields, like ripples on the surface of an ocean.

Quantum Fluctuation: A temporary change in the amount of energy at a point in space, as the uncertainty principle allows. These fluctuations explain the Casimir effect and may have triggered the Big Bang.

Quantum Number: A value that describes a specific property of a quantum system, such as energy level, angular momentum, or spin. Quantum numbers determine what states are possible for particles in atoms.

Quantum Superposition: The ability of a quantum system to exist in multiple states simultaneously until measured. This principle underlies phenomena like quantum computation and quantum tunneling.

Quantum Teleportation: A technique using quantum entanglement to transmit quantum information from one location to another. Despite its name, it doesn't transport matter or energy, only quantum states.

Quantum Tunneling: A phenomenon where particles pass through energy barriers that they classically couldn't overcome. This effect explains radioactive decay and makes specific electronic devices possible.

Qubit: The quantum equivalent of a classical computer bit. Unlike classical bits that are either 0 or 1, qubits can exist simultaneously in a superposition of both states, enabling quantum computation.

Schrödinger Equation: The fundamental equation of quantum mechanics that describes how quantum states evolve. It plays a role similar to Newton's laws in classical mechanics.

Spin: A quantum property of particles that has no classical analog. Despite its name, it's not about physical rotation. Still, it is instead an intrinsic form of angular momentum that can only take specific discrete values.

String Theory: A theoretical framework attempting to unify quantum mechanics with gravity by proposing that all particles and forces are different vibrations of tiny one-dimensional "strings."

Vacuum Energy: The energy present in space due to quantum fluctuations. This energy might be related to dark energy and the universe's expansion.

Wave Function: The mathematical description of a quantum system that contains all possible information about its state. The square of its amplitude gives the probability of measuring particular properties.

Wave-Particle Duality: The principle that all matter and energy exhibit wave-like and particle-like properties, demonstrating fundamentally different behavior depending on how they're observed or measured.

This completes our comprehensive glossary of quantum physics terms. These concepts form the foundation for understanding the strange quantum world and its applications in modern technology. While some terms might seem abstract, remember that they describe natural phenomena that scientists observe and use daily in laboratories.

Recommended Bundle

Suppose you've enjoyed exploring the quantum nature of reality in this book. In that case, you might wonder about even more profound questions about the nature of our universe. Are we living in a simulation? Could multiple universes exist simultaneously? What lies beyond our observable reality?

The "Exploring Simulated Worlds & Parallel Realities" bundle complements your quantum physics journey. Just as we've explored how quantum mechanics challenges our everyday assumptions about reality, this bundle delves into equally mind-bending possibilities about the fundamental nature of existence.

You'll discover how the simulation hypothesis relates to quantum computation, how the many-worlds interpretation of quantum mechanics might connect to the multiverse theory, and how our understanding of consciousness might bridge these various perspectives on reality. Written with the same clarity and accessibility as this quantum physics guide, this bundle provides the perfect next step in your exploration of nature's most profound mysteries.

Scan the QR code below to begin your journey into simulated worlds and parallel realities. This will take you directly to the Amazon page to purchase the complete bundle.

Scan the QR Code to Explore 'Simulated Worlds & Parallel Realities' on Amazon

Scan the QR code to purchase the complete "Exploring Simulated Worlds & Parallel Realities" bundle on Amazon. Your next adventure into the nature of reality is just one quantum leap away.

Entangle Others in the Wonder of Quantum Physics

Did this book help demystify quantum physics for you? Share your quantum leap in understanding by leaving a review! Your feedback helps other curious minds discover that quantum physics isn't just for rocket scientists. Plus, it helps us know what clicked (or what left you uncertain) so we can make science more accessible for everyone.

Whether you're now confidently explaining superposition over dinner or just starting to grasp why Schrödinger's cat is both alive and dead, your experience matters. Take a moment to share your thoughts - it's the observer effect in action! Your review could be why someone else decides to explore the quantum realm.

Leave your review wherever you purchased the book or on your favorite reading platform. Thank you for being part of our quantum community!

www.ingramcontent.com/pod-product-compliance
Lightning Source LLC
Chambersburg PA
CBHW071431210326
41597CB00020B/3738

* 9 7 8 1 9 6 2 2 5 4 1 5 1 *